The Road to Azure Cost Governance

Techniques to tame your monthly Azure bill with a continuous optimization journey for your apps

Paola E. Annis

Giuliano Caglio

BIRMINGHAM—MUMBAI

The Road to Azure Cost Governance

Group Product Manager: Aaron Lazar
Publishing Product Manager: Sathyanarayanan Ellapulli
Senior Editor: Rohit Singh
Content Development Editor: Rosal Colaco
Technical Editor: Maran Fernandes
Copy Editor: Safis Editing
Project Coordinator: Ajesh Devavaram
Proofreader: Safis Editing
Indexer: Manju Arasan
Production Designer: Sinhayna Bais
Marketing Coordinator: Sonakshi Bubbar

First published: December 2021
Production reference: 1140122

Published by Packt Publishing Ltd.
Livery Place
35 Livery Street
Birmingham
B3 2PB, UK.

ISBN 978-1-80324-644-4

www.packt.com

Foreword

For companies that have started or planned a Digital Transformation program, cloud adoption is a must rather than an option. The flexibility, scalability, and resilience of cloud computing enable a reduced time to market, faster technology adoption, and manageability and sustainability of solutions, which are essential values for a modern company.

During my career, I've faced different approaches to the cloud, from *modernize your application first* to *lift and shift everything*, and I've also had the opportunity to challenge my teams on every different aspect of the cloud adoption mission.

Migrating your modern applications may slow down cloud adoption, and while lifting and shifting may be faster, you'll soon realize that lifting and shifting is only the beginning of the journey to cloud adoption. This book will help you focus on the right journey according to your adoption maturity, so that it's affordable for you.

In my current company, we chose a complete *lift and shift approach* in 2018, switching from an on-premises data center to a full public cloud. We quickly had to learn the first lesson: public cloud requires a different mentality and culture that legacy companies were not yet ready to embrace. And so, in this book, we have dived into this and worked hard to understand, govern, and optimize our cloud services.

Soon we became aware of the second lesson: we need to evolve our services cloud costs understanding. We also need to advance through the infrastructure-based cost representation to the business application one and take advantage of our already present application catalog and documentation. This kind of evolution is perfectly described in this book, and it has permeated our cloud governance model so much that it led us to share cost and performance ownership between the people contributing to the infrastructure and the application owners. Using the end-to-end cost view also allowed us to rewrite the architectural and application guidelines to make our infrastructure increasingly more effective and efficient.

Time goes by, and today I can safely say we are at yet another crossroads. Here, we need to decide how legacy applications will need to either be decommissioned or refactored into cloud-native architectures, and our governance process is the foundation of how we'll be able to play and win this game. This is why I can truly relate to all the topics covered in this book with my experience.

After several years of cloud adoption and its evolution, governance, and control, I can candidly say that adopting a public cloud is not all pretty nor easy, but it's essential. The reward for the braves is a speed of innovation that was unimaginable only a few years ago. You'll also give your company the chance to be a leader in digital transformation, and to be a business that stands the test of time having completed its cloud adoption mission.

In this book, Paola and Giuliano lead us in understanding the cloud world by offering the tools to prepare, face, and manage a challenge that while complex, boasts great benefits for IT and Business.

Marco Barra Caracciolo

Chairman and CEO at Bludigit SpA

Contributors

About the authors

Paola E. Annis has worked for over 25 years in IT. Currently at Microsoft enjoying the Azure cloud, she has extensive experience working with strategic customers on large migrations and digital transformation projects. In her daily work, she advocates cost governance, sustainable software engineering, and women in tech communities. She lives in Milan with her kids, cats, and husband. In her spare time, she enjoys heavy metal music while embarking on improbable DIY projects for her country house.

> *I want to thank my family and all my geek friends who have inspired me and supported me through the years and technologies (you know who you are). Special thanks to, Davide Bedin, Fausto Massa, and everyone else at Microsoft who helped me fall in love with Azure.*

Giuliano Caglio has over 15 years of experience. He is constantly looking for challenges in infrastructure management to relate to applications and infrastructure and thus reach a broader view of cloud environments, asset management, and cost governance. He lives near Milan with one daughter, two turtles, and his wife. In his spare time, he enjoys gardening, electronics, and railway modeling DIY projects.

> *I want to thank my cloud team for supporting me in evolving the cost control model and dashboard, and Alessandro Santandrea and Domenico D'Amore for challenging me on business reporting requirements and solutions. Special thanks to Matteo Borgonovo and Christian Meggiorin for supporting and putting up with me while working on challenging projects.*

> *We would also like to thank the whole team at Packt, and the technical reviewers for helping and guiding us throughout this adventure and for making it fun!*

About the reviewers

Wesley Haakman started working in IT at the age of 17, providing support to end users in a health care organization. Over the next couple of years, he moved through different roles and went from managing and maintaining the IT infrastructure to leading the IT team. After a couple of years, he took a deep dive into the world of Azure and joined Intercept. After 8 years, he now works as a lead Azure architect for Intercept and spends his days designing, implementing, and optimizing customer environments, as well as providing workshops on behalf of Intercept. Additionally, he is a Microsoft Azure MVP and is always happy to provide a session at an event or user group, or just join a meeting over Microsoft Teams to talk Azure.

Bruce Cosden has a broad background in supporting enterprise customers' technology transformation journeys, as well as a life-long deep personal interest in technology and its impact on society. A computer science major, Bruce has created, produced, and managed the creation of highly reliable enterprise applications for Fortune 100 companies in financial services and the media and entertainment space. Bruce has over 15 years of experience as a management consulting technology leader and as a cloud solution architect at Microsoft. Currently, Bruce is an advanced networking specialist at Microsoft.

Arvind Kumar Soni is a senior Azure cloud architect with over 21 years of extensive experience in Microsoft technology. He serves as an advisory architect focusing on enterprise architecture, application modernization and migration, cloud governance, and cost management. He has certifications including TOGAF, Azure Solution Architect, Azure DevOps, Azure Security, and IBM Cloud Technical Advisor.

Atul J. Kamble is the founder and CEO of Cloudnautic. Atul helps organizations to engage IT transformation and cloud solutions since 2016. Atul has over 5 years of experience in providing IT and cloud solutions. He received his master of technology in information technology from Walchand College of Engineering, Sangli, India, in 2018 and his bachelor of engineering in computer engineering from the Government College of Engineering and Research, Pune, India, in 2016.

Table of Contents

Section 2: Cloud Cost Savings

4
Planning for Cost Savings – Right-Sizing

5

Planning for Cost Savings – Cleanup

6

Planning for Cost Savings – Reservations

Section 3: Cost- and Carbon-Aware Cloud Architectures

7

Application Performance and Cloud Cost

8

Sustainable Applications and Architectural Patterns

Assessments

Index

Other Books You May Enjoy

Preface

In my experience as a cloud architect in the past few years, I worked with many customers – companies from small to large, public and private, innovative and conservative. Each of them had their own idea in mind of what the cloud could mean for their business, which is the right approach, but none of them knew for certain what to do once they got in the cloud.

One of the most difficult conversations with a customer approaching the cloud revolves around how this new technology is changing some of the roles of IT, sometimes through simple adaptive learning by acquiring the new technology skills, and sometimes through a tough transformation by adding roles that a company didn't need or know about before, such as the role of buying cloud services. This is the case with the cloud cost manager, the owner of cloud billing. It's a very specific role that in the past might have resembled that of a finance controller, but today requires technical skills to understand metering, automation, and services, and especially to know what is essential to the IT core and what can be switched off or swapped.

If you are reading this book and have been wondering who takes care of cloud spending, it is a new breed of technical experts. They are not only very skilled in IT infrastructure, design patterns, frameworks, and architectures, and capable of understanding cloud trends and innovation, but are also fluent in finance, costs, and meters, understand the principles of cloud services, and are able to sustain a tough conversation with the company's finance team and sometimes the CEO. In this book, we will learn about the Azure cloud billing process and how to govern that process with your own optimization patterns and automation, so that surprises in the monthly bill are reduced to a minimum and cloud spending becomes part of the company's finance operations with ease.

In the first section of the book, we'll start by learning how to read the monthly bill and understand all the monitoring and control options that Azure offers, along with how to organize your cloud resources to match your company's spending hierarchy. The second section is dedicated to cost-saving techniques: from right-sizing your workloads (either Azure IaaS or PaaS) to cleaning up unused services, to the reservations process. The final section is dedicated to optimizing your application and database with the goal of reducing infrastructural costs.

The purpose of this book is to familiarize yourself with Azure cloud billing and all the steps needed to embrace the public cloud spending with new governance and continuous optimization processes that will bring clarity and governance across all the company's departments and applications.

Who this book is for

If you are someone who deals with Azure cloud costs and have a technical background, this book will help you understand and control your cloud spending. Created with the logic and experience of various large organizations dealing with this cloud spending struggle, this book is for decision-makers, cloud managers, cloud architects, cost controllers, and software solution professionals working with Microsoft cloud services in Azure and looking to build optimized solutions for their enterprise operations.

What this book covers

Chapter 1, Understanding Cloud Bills, enables you to gather insights, audit, and enforce automation and policies for successful cloud cost governance. You will also gain a deep understanding of how service categories and cloud meters work.

Chapter 2, What Does Your Cloud Spending Look Like?, provides you with several tools and guidelines to successfully implement structured cost governance that matches the financial and technical KPIs of your company.

Chapter 3, Monitoring Costs, enables you to put in place a cost administration process and successfully monitor and control costs through dashboards and alerts.

Chapter 4, Planning for Cost Savings – Right-Sizing, will help you understand and use cost-saving tools such as reservations, along with techniques to downsize, switch off and on, and offer on-demand solutions, where possible, for Azure resources.

Chapter 5, Planning for Cost Savings – Cleanup, helps you learn how to properly set up a control process to handle reservations: from understanding the model to successfully driving a full capacity strategy of reserved services.

Chapter 6, Planning for Cost Savings – Reservations, will help you get familiar with a clear path and process of handling services and Reservations in Azure, from the initial purchase process to the daily utilization check and periodic changes that might be needed.

Chapter 7, Application Performance and Cloud Cost, helps you learn how a well-designed and performant application can bring down cloud costs and shows how an investment in performance optimization strategies, such as database tuning, is easily and quickly repaid by the monthly savings.

Chapter 8, Sustainable Applications and Architectural Patterns, focuses on the new concept of sustainable (or green) software engineering and how applications designed for performance and cost management can also have a lower impact on their carbon footprint.

Assessments contains all the answers to the questions in all the chapters of this book.

To get the most out of this book

To get the most out of this book you should have the following:

1. Access and privileges to the Azure portal with reader role (at a minimum) for the Azure subscription(s) you want to control..

2. Privileges to see cloud costs for your Azure subscription(s).

3. Access to a CLI (either PowerShell, or Cloud Shell via the Azure portal).

4. An open mindset to understand, manage, and squeeze fluid recurring entities such as the public cloud costs.

Software/hardware covered in the book	Operating system requirements
Azure Cloud Shell	Windows, macOS, or Linux
Azure Cloud Services	Windows, macOS, or Linux
PHP 7.4 and above	Windows, macOS, or Linux

You can find out how to get access to the Azure portal and the necessary privileges to access Cost Management information here: `https://docs.microsoft.com/en-us/azure/cost-management-billing/costs/assign-access-acm-data`.

If you are using the digital version of this book, we advise you to type the code yourself or access the code from the book's GitHub repository (a link is available in the next section). Doing so will help you avoid any potential errors related to the copying and pasting of code.

Download the example code files

You can download the example code files for this book from GitHub at `https://github.com/PacktPublishing/The-Road-to-Azure-Cost-Governance`. If there's an update to the code, it will be updated in the GitHub repository.

We also have other code bundles from our rich catalog of books and videos available at `https://github.com/PacktPublishing/`. Check them out!

Download the color images

We also provide a PDF file that has color images of the screenshots and diagrams used in this book. You can download it here: `https://static.packt-cdn.com/downloads/9781803246444_ColorImages.pdf`.

Conventions used

There are a number of text conventions used throughout this book.

`Code in text`: Indicates code words in text, database table names, folder names, filenames, file extensions, pathnames, dummy URLs, user input, and Twitter handles. Here is an example: "Once you've created the empty structure, you could use the second script, from the `ingest.php` file, to load the `.csv` file from the Azure Cost Management tool."

A block of code is set as follows:

```
Tag name  :  BsnApp
Tag value :  |Application1234|
Tag name  :  Landscape
Tag value :  Production
```

Any command-line input or output is written as follows:

```
$ mkdir css
$ cd css
```

Bold: Indicates a new term, an important word, or words that you see onscreen. For instance, words in menus or dialog boxes appear in **bold**. Here is an example: "In your Azure portal, navigate to **Cost Management + Billing | Cost Management** and select **Budgets**."

> **Tips or Important Notes**
> Appear like this.

Get in touch

Feedback from our readers is always welcome.

General feedback: If you have questions about any aspect of this book, email us at customercare@packtpub.com and mention the book title in the subject of your message.

Errata: Although we have taken every care to ensure the accuracy of our content, mistakes do happen. If you have found a mistake in this book, we would be grateful if you would report this to us. Please visit www.packtpub.com/support/errata and fill in the form.

Piracy: If you come across any illegal copies of our works in any form on the internet, we would be grateful if you would provide us with the location address or website name. Please contact us at copyright@packt.com with a link to the material.

If you are interested in becoming an author: If there is a topic that you have expertise in and you are interested in either writing or contributing to a book, please visit authors.packtpub.com.

Share Your Thoughts

Once you've read The Road to Azure Cost Governance, we'd love to hear your thoughts! Scan the QR code below to go straight to the Amazon review page for this book and share your feedback.

https://packt.link/r/1-803-24644-8

Your review is important to us and the tech community and will help us make sure we're delivering excellent quality content.

Section 1: Cloud Cost Management

In this section, you will gain a full understanding of the Azure billing model and how to put a cost governance model in place with one or more tools for monitoring and controlling costs. To be able to audit and enforce automation and policies for successful cloud cost governance, it is necessary to have a deep understanding of how service categories and cloud meters work.

We'll learn several tools and guidelines to successfully implement a structured cost governance model that matches the financial and technical KPIs of your company. At the end of this section, you will have a cost governance process in place, and will be able to successfully monitor and control costs through dashboards and alerts.

This section comprises the following chapters:

- *Chapter 1, Understanding Cloud Bills*
- *Chapter 2, What Does Your Cloud Spending Look Like?*
- *Chapter 3, Monitoring Costs*

1

Understanding Cloud Bills

In this first chapter of the book, we will guide you so that you are able to visualize, export, and understand Azure billing from the common Azure portal. We will cover the following topics in the chapter:

- So, you're doing cloud—now what?
- Understanding how cloud billing works
- Reviewing the Azure portal for cost management
- Matching your cloud billing with your company's department organization
- Sources of billing information and export methods of billing data

By the end of this chapter, you will have a deep understanding of how service categories and cloud meters work, and be able to assess, evaluate, and export the cloud billing of your managed Azure subscription(s).

Technical requirements

To follow this chapter, you'll need the following:

- A computer with internet connectivity

- Access to the Azure portal

- Privileges to access Cost Management information (`https://docs.microsoft.com/en-us/azure/cost-management-billing/costs/assign-access-acm-data`)

You can find the source code used in this book here: `https://github.com/PacktPublishing/The-Road-to-Azure-Cost-Governance`

So, you're doing cloud—now what?

The public cloud is becoming a mainstream technology staple for every modern company, and most customers have large, skilled, and experienced **information technology** (**IT**) teams that are converting from managing on-premises resources to hybrid or pure cloud.

One of the most neglected aspects of public clouds, from an IT department's point of view, is how to deal with recurrent, granular, and detailed cloud costs that are in no way associated with the old purchase process of on-premises resources and require technical skills to be fully understood.

Migrating entire data centers to the cloud is not the arrival point: it is, in fact, only the starting point of a journey toward a pay-as-you-go model, where technology changes along with mentality, ways of working (think of **development-operations**, more commonly known as **DevOps**), and flexibility all have a direct, daily, mutable impact on your IT spend, unlike anything you have ever managed before. The shift from **capital expenditures** (**CapEx**) on-premises to **operational expenditures** (**OpEx**) on the cloud is a major game-changer for every company, and—specifically—per meter billing is a challenge when moving from a fixed cost pattern to a variable and more fluid cost.

This chapter is dedicated to how Azure bills services and cloud objects, and how to retrieve all the necessary information to then proceed with the next step, which is choosing your own cloud spending strategy. The goal is for you to fully understand cloud billing and all its implications, from technical to financial.

Understanding how cloud billing works

This section will guide you through the most common patterns of Azure cloud billing, to be able to correctly display, analyze, and export your billing information and associate it with the relevant IT resources.

In Azure, once you have access to a portal and have privileges to create resources and objects, the billing meters start, according to a few concepts that we'll describe in this chapter, as outlined here:

- **Type of agreement with Microsoft (Enterprise Agreement (EA), Microsoft Customer Agreement (MCA), cloud solution provider (CSP)**, developers, and more): This is a contract signed with Microsoft that will allow you to create Azure subscriptions and cloud services within those subscriptions.

- **Services' price**: The price of Azure services, along with any discount or promotion or benefit applied to services and or contracts.

- **Resource utilization (region, time, frequency, and type)**: This is a key concept of cloud services. Every cloud service has a different pricing scheme that depends on these variables. For example, a **Virtual Machine (VM)** is billed by the duration of utilization and according to the VM family and size, and it's also different for each selected region. Therefore, the final price will be defined by how much time a specific VM type in an Azure region (that is, West US) has been running. We'll be able to see practical examples in the pricing section.

The way you can use resources in Azure depends on what type of agreement you have with Microsoft, even if in the end all will be visualized through the Azure portal.

At the time of writing, there are a few different types of billing accounts, as outlined here:

- **Microsoft Online Services Program (MOSP)**: This type of contract can be created by signing up through the Azure website. Each subscription will have its own payment methods and monthly invoice, as shown here:

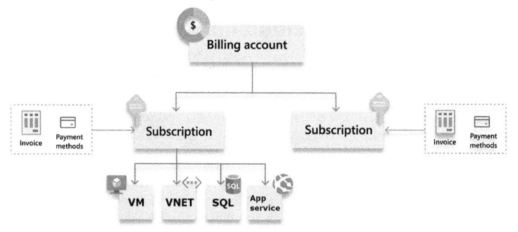

Figure 1.1 – Structure of the MOSP

- **EA**: This is created when your company signs an EA with Microsoft. Each agreement will have one or more enrollments where Azure subscriptions can be created. Enrollment is typically associated with a monetary commitment or pre-paid agreements that grant special discounts to companies (the agreement may drop if the customer is not consuming as promised, whereas on exceeding consumption they will get the same discounts). The invoice will be unique across the whole EA, and the payment method is defined within the EA. The structure for an EA billing account is shown in the following diagram:

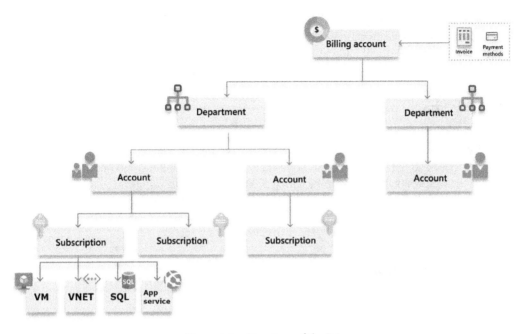

Figure 1.2 – Structure of the EA

- **Microsoft Customer Agreement (MCA)**: This is a customer agreement only for Azure resources, with a limit of 20 subscriptions for each profile. A diagram of this agreement is shown here:

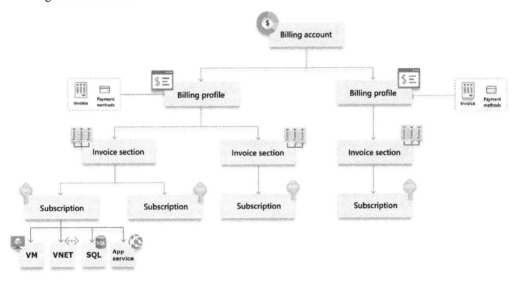

Figure 1.3 – Structure of the MCA

- **Microsoft Partner Agreement** (**MPA**): This agreement is specific to partners (CSP) so that they can manage billing for their customers. If your Azure solution is through a CSP, you will have to request specific access to be able to see the billing. In addition, please be aware that Cost Management access will display retail rates, therefore you will need to make adjustments to reflect any additional discounts or benefits for your contract.

> **Important Note:**
>
> When a customer works with the MPA, they don't own the billing account, and all the features that come with the billing account are only accessible through the CSP itself. Tasks such as creating subscriptions need to be done through the partner, as well as the responsibility of providing such features through a **Cost Management Platform** (**CMP**), as for the cloud billing access.

The following diagram shows the structure of the MPA:

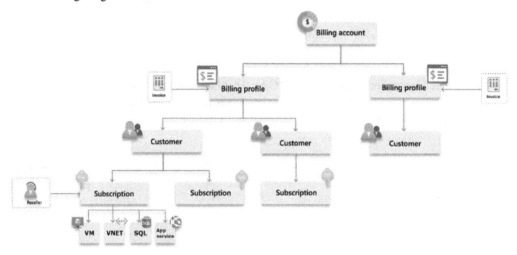

Figure 1.4 – Structure of the MPA

> **Tip:**
>
> A full list of supported Microsoft Azure offers can be found here: `https://docs.microsoft.com/en-us/azure/cost-management-billing/costs/understand-cost-mgt-data`. Other offer types unsupported by **Azure Cost Management** can still benefit from this book's suggestions, provided that they can find a way of exporting their cost details.

Cost Management information can be accessed by global admin(s) and admin agents and will display the invoiced pricing for Microsoft partners and providers. This is typically then used to charge back Azure usage to customers and resellers and to view specific billing benefits (such as Partner Credit), user budgets, exports, and notifications.

End customers and resellers can only view their specific subscription(s) billing information with proper **Role-Based Access Control** (**RBAC**) access to the subscription(s) and the visibility policy enabled for the tenant: the display information can only be at retail prices.

Additional information can be found here: `https://docs.microsoft.com/ en-us/azure/cost-management-billing/costs/get-started-partners`.

Question 1: *How do I know what my billing type is?*

As per the documentation at `https://docs.microsoft.com/en-us/azure/ cost-management-billing/manage/view-all-accounts#check-the- type-of-your-account`, you can check your type of billing directly from the Azure portal's **Azure Cost Management** | **Properties** page.

Each of these billing types allows you to create Azure subscriptions, which are the largest technical repositories of resources in Azure.

You can then access billing information via several portal pages and dashboards according to the scopes you choose, as shown in the next screenshot:

Figure 1.5 – Azure portal Cost Management + Billing page

The Azure portal **Cost Management + Billing** page will display a table with all the available scopes according to the following:

- Your credentials
- Your access permissions (**identity and access management**, or **IAM**)
- Your company's hierarchy

For Azure plans that allow invoices to be read and downloaded, this can be done via the Azure portal through the following:

- MOSP
- MCA
- MPA

Question 2: *How are resources billed in Azure?*

Well, the truth is that it depends on the resource and the meter category and subcategory. Also, considering the whole set of data center services, not all resources are billed: for example, a **network interface card** (**NIC**) attached to a VM is typically not billed.

Every Azure service that is billed has a metering unit that will define how the object can be considered in terms of a paid service—for example, a VM is billed according to its capacity (the VM size is the meter subcategory) and its usage frequency (VMs are billed per duration of usage). In the final monthly bill or invoice, every used Meter Category and subcategory is summed according to its usage and pricing.

Platform-as-a-Service (**PaaS**) services are typically billed by performance tiers—such as standard, premium, basic, and so on—depending on the type of service and the combination of resources that are included in each tier.

Let's review a few examples of how various Azure services have slightly different billing rules, as follows:

- **VMs**: Azure VMs are billed per second rounded down to the last minute, but the different VM states (if the VM is starting, or doing **operating system** (**OS**) provisioning, and so on) will help you decide whether or not to pay for the VM, as explained in the following screenshot. This is important when you are switching off a VM for cost reduction (we'll learn more about this in the upcoming chapters) and must make sure you deallocate it in order to stop its billing (additional information can be found here: `https://docs.microsoft.com/en-us/azure/virtual-machines/states-billing`):

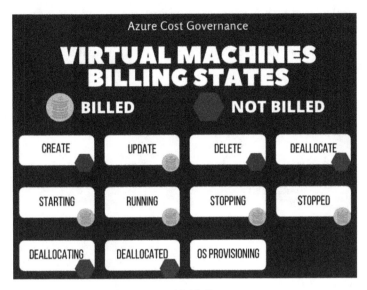

Figure 1.6 – VM billing states

- **App Service** (according to the official pricing given at `https://azure.microsoft.com/en-us/pricing/details/app-service/windows/`) has six different pricing plans, each with a different set of features, **CPU**/memory, and pricing:

  - Free
  - Shared
  - Basic (B1, B2, B3)
  - Standard (S1, S2, S3)
  - Premium (P1V2, P2V2, P3V2, P1V3, P2V3, P3V3)
  - Isolated (I1, I2, I3, I1V2, I2V2, I3V2, I1V3, I2V3, I3V3)

The price is calculated by hours of usage, and users can add items on top of the plan, such as domain, certificate, and **Secure Sockets Layer (SSL)** connections. The following screenshot summarizes all the App Service available tiers and the related included features:

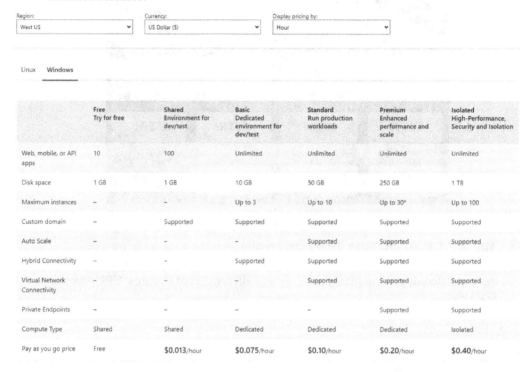

	Free Try for free	Shared Environment for dev/test	Basic Dedicated environment for dev/test	Standard Run production workloads	Premium Enhanced performance and scale	Isolated High-Performance, Security and Isolation
Web, mobile, or API apps	10	100	Unlimited	Unlimited	Unlimited	Unlimited
Disk space	1 GB	1 GB	10 GB	50 GB	250 GB	1 TB
Maximum instances	–	–	Up to 3	Up to 10	Up to 30*	Up to 100
Custom domain	–	Supported	Supported	Supported	Supported	Supported
Auto Scale	–	–	–	Supported	Supported	Supported
Hybrid Connectivity	–	–	Supported	Supported	Supported	Supported
Virtual Network Connectivity	–	–	–	Supported	Supported	Supported
Private Endpoints	–	–	–	–	Supported	Supported
Compute Type	Shared	Shared	Dedicated	Dedicated	Dedicated	Isolated
Pay as you go price	Free	$0.013/hour	$0.075/hour	$0.10/hour	$0.20/hour	$0.40/hour

Figure 1.7 – App Service plans

- Other examples of pay per use are **Cosmos DB** and **Functions**. Cosmos DB (https://azure.microsoft.com/en-us/pricing/details/cosmos-db/) normalizes all database operations and offers two database models, as follows:

  - Provisioned throughput (measuring **Request Units (RUs)**)
  - Serverless (billing RU)

  On top of it, storage must be added.

For the serverless model, the following screenshot displays the Cosmos DB pricing for RUs:

Serverless	Total Request Units (RU)	Price per 1M RU
Serverless request units (RU)	1,000,000	$0.279
Serverless request units (RU) with availability zone	1,000,000 x 1.25	$0.279

Figure 1.8 – Cosmos DB pricing for RUs

The next screenshot shows the Cosmos DB pricing for storage consumed:

Consumed Storage	Total GB	Price
Transactional storage (row-oriented)	1 GB	$0.25/month
Analytical storage (column-oriented)	1 GB	$0.03/month

Figure 1.9 – Cosmos DB pricing for storage

In addition, users must take into account backup storage, analytical storage, dedicated gateway, and multiple regions. All these configurations should be evaluated during the project design phase to better understand the impact on costs that a specific choice will have according to how that specific resource is billed.

According to its pricing rules (mentioned at `https://azure.microsoft.com/en-us/pricing/details/functions/`), with Azure Functions, the pay-per-use model has a number of free grants (1 million requests and 400,000 **gigabytes (GB)** of resource consumption) per month, and a price per execution and execution time, as displayed in the following screenshot:

Meter	Price	Free Grant (Per Month)
Execution Time*	$0.000016/GB-s	400,000 GB-s
Total Executions*	$0.20 per million executions	1 million executions

Figure 1.10 – Azure Functions pay-per-use price

There is also a premium plan with reserved **virtual CPU (vCPU)** and memory, as illustrated in the following screenshot:

Meter	Price
vCPU duration	vCPU: **$0.17** vCPU/hour
Memory duration	Memory: **$0.0122** GB/hour

Figure 1.11 – Azure Functions Premium plan

Most of the billing categories can be summarized by service, such as Compute, Storage, Networking, Web, Databases, Identity, Security, Media, and so on, and have their own timeframe, unit, and frequency of billing metering.

Question 3: *What are MACC and monetary commitment?*

You might have signed with Microsoft a **Microsoft Azure Consumption Commitment (MACC)**, a contractual agreement that upon a commitment of Azure spend, they will grant you special discounts. If your company has a MACC, it is important that you track your monthly and yearly spending against the commitment on a recurring basis, together with all other cost governance matters. This is mainly because your commitment to consuming Azure services will have an impact on the discounts and benefits from Microsoft, and its unfulfillment may have an impact on your grants.

Question 4: *How do I know the cost of my Azure services?*

You might have got pricing information from your partner, your Microsoft account team, or other sources of information, but for the sake of speed and comprehension, we recommend that you become fluent with using the Azure pricing calculator. This will allow for the following:

- A better understanding of the billing metering for each service
- An updated spending view of new objects in Azure
- A clear spending forecast for new projects

The Azure pricing calculator can be found at `https://azure.microsoft.com/en-us/pricing/calculator/`.

Here is an example of how you can price a service. Let's imagine you have a very simple application that is comprised of a couple of frontend VMs and a couple of backend VMs. The requirements are very basic, as we can see here:

- 4 cores; 16 GB **RAM** for the frontend VMs

- 8 cores; 32 GB RAM for the backend VMs

The first pricing example is for frontend VMs. You choose the OS type and VM family, and enter the details on the configurator's page, as illustrated in the following screenshot:

Figure 1.12 – Adding a VM to the Azure pricing tool

At this point, you need to know what type of usage these VMs will endure. You can choose from the following:

- **Always on** (maybe they are production VMs). In this case, 730 hours is the full month of an always-on VM. If this is the case, you will probably want to consider reservations, which we'll be able to dig deeper into in the next chapter, but for the sake of pricing, you might want to decide on a 1-year or 3-year reservation, to lower the VM price, as illustrated in the following screenshot:

Savings Options

Save up to 72% on pay-as-you-go prices with 1-year or 3-year Reserved Virtual Machine Instances. Reserved Instances are great for applications with steady-state usage and applications that require reserved capacity. Learn more about Reserved VM Instances pricing.

Compute (D4 v4)

⦿ Pay as you go
○ 1 year reserved (~41% discount)
○ 3 year reserved (~62% discount)

Figure 1.13 – Azure VM pricing: choice of pay as you go or reservation options

- A **schedule interval**—for example, Mon-Fri, 8 a.m.-8 p.m., which amounts to 12 hours, 5 days per week, for a total of 240 hours per month

- An **on-demand VM**, typical of development environments, where developers or workers will switch the VM on only when using it—for example, 1 week per month, 12 hours per day, for a total of 60 hours per month

If you change the billing hours of the chosen VM, it will become clear that deciding on the usage time has a strong impact on billing at the end of the month! We'll be able to discover ways of optimizing this choice in the next chapters. The VM pricing lets you add storage options for the OS, along with any storage transactions related to Standard **HDDs**, as illustrated in the following screenshot:

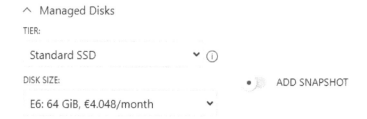

Figure 1.14 – Azure VM price: choice of managed disks

The tool will also help you in calculating data transfer charges where applicable, as shown in the following screenshot:

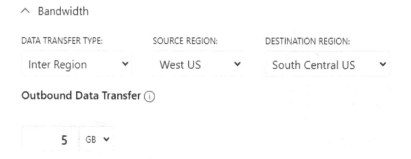

Figure 1.15 – Azure VM price: bandwidth options

You will then need to add any additional data disks via storage service pricing, and even with storage you will have the option of reserving capacity—we will talk about this in the next chapter. This is illustrated in the following screenshot:

Figure 1.16 – Azure storage account pricing for additional data disks

In this section, we started with explaining the subscription hierarchy and billing of Azure, according to the contract in place with Microsoft, and then provided a brief explanation of the Azure pricing calculator, which is a very helpful tool not only to understand new project costs but also to analyze the impact of savings once the cost optimization process is in place.

Once you have organized your company's subscriptions and have access to each of them, including—where applicable—the root management group (we will work on management groups and privileges later in the book, but for now, the important information is that the cost owner should have full access to all cost-related objects in Azure), it is time for you to open the Azure portal for cost management and start looking at the billing, as we will see in the next section.

Reviewing the Azure portal for cost management

This section will guide you through the Azure portal's **Cost Management + Billing** section. The first thing to know is that billing is reflected with up to 24 hours' lag in the portal, which means that the spending displayed on the portal has roughly a 1-day delay.

When you open the **Cost Management + Billing** page, you are prompted with **Billing** scopes, **Cost Management**, and management groups links. When you click on the **Cost Management** link, you are prompted with the menu shown in the following screenshot (at the time of writing, there were a few interesting pages in the Azure portal's **Cost Management** section):

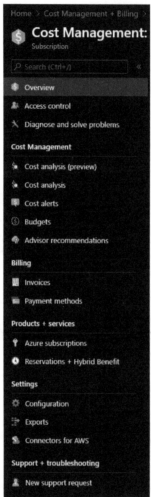

Figure 1.17 – The Azure portal Cost Management view pane

The Azure portal **Cost Management** section features a few very useful pages that will allow you to review billing and charges for the subscriptions and agreements where you have the correct privileges (either **Owner**, **Contributor**, **Reader**, or any custom role with access to costs). For partner agreements, you might need to ask your reseller to grant you access to download the billing information for you.

Azure customers with an Azure EA, MCA, or MPA may view and download their pricing in the Azure portal. Depending on the policies set for your organization by the Enterprise admin, only certain administrative roles provide access to your organization's EA pricing information. If you have an MCA, you must be the billing profile owner, contributor, reader, or invoice manager to view and download pricing. If you have an MPA, you must have the **Global Admin** and **Admin Agent** role in the partner organization to view and download pricing.

Matching your cloud billing with your company's department organization

The first step to cost governance is to understand the hierarchy of subscriptions and how this can be adapted to the hierarchy of your organization, or vice versa. The goal of this section is to help you define a structural organization for billing that matches the company's one: we will be able to put this into practice in the next chapter by setting up automation and configuration dedicated to billing.

There are several ways to organize subscriptions, and for larger organizations, the **Management Groups** feature helps manage large numbers and complex hierarchies of subscriptions, in terms of cost, security, and governance. Blueprints help define in a declarative way how you intend to organize and set up your environment.

Defining a subscriptions and scopes hierarchy will allow you to match the billing and chargeback expectations of your company for optimal cost control.

There are several ways to do this, but the choice of hierarchy has a few technical implications that might make it difficult to change afterward, which is why Microsoft recommends carrying out this operation within the Well-Architected Framework and Cloud Adoption Framework governance models so that the landing zone will also consider the billing necessities of the finance and other departments.

You might decide to organize your subscriptions in the following way:

- By department/business unit (that is, marketing, IT, **HR**, and so on)
- By function (shared services, data services, security, and so on)
- By geography (either by continent, region, or the geographic subsets specific to your company)
- By application/workload

Or, you could use a mix of all the preceding options, especially when considering an already deployed virtual data center where you are reorganizing mixed and stratified configurations.

In my experience, large organizations have very structured IT roles that might clash with fast-paced, constantly shifting cloud roles. *A cloud architect is supposed to work on a subscription, but can they also be the owner of such a subscription?* A **financial operations** (**FinOps**) team may need to have technical skills to understand cloud billing while owning subscriptions and accounts. *When a new project starts, should the owner of the project also own the subscription or just be a contributor?* Department owners are typically the first stop when looking for subscription owners, but in some companies, there is no chargeback for cloud services, and in the end, this might be useless.

The main question you should pose is: *How do I want to organize the cloud spending of my company according to its internal processes and departments?* And this conversation should happen as soon as possible. In the next chapter, we'll be able to implement several types of hierarchy for subscriptions, using special features such as management groups.

The takeaway of this section is that your technical department will have to work with the cloud cost owner to understand how the virtual data center must be deployed to *make it easier for the finance department to understand and process billing information.* This might seem a lower-priority task compared to the technical deployment of objects and resources, but if you don't set the correct expectations before any deployment happens, you may find it very difficult to be able to track costs according to the department or application that is using those resources.

Sources of billing information and export methods of billing data

In this section, we will go through several basic concepts of cloud cost representations and ways of getting cost information, with the related pros and cons.

As mentioned, before jumping into each of the recommended ways of checking your cloud spending, you need to take a little time to clarify what are the targets you want to achieve (it's very useful to also involve the financial controller team of your company in this).

You could start with some of the main questions you typically have to answer, as follows:

Question 5: *Do you need a technical representation of the costs, a business representation, or both?*

A technical representation is easily achievable since it is based on the infrastructure you've built on Azure and represents the costs for a technical team. A business view of the costs should link the costs to a *business object* (for example, an application or a service to the customer) that may be composed of different Azure resources. Tagging cloud resources is an important part of this representation and will be covered in the upcoming chapters.

Long story short, you may think about the technical representation as *how much we spent on storage, grouped by storage type and storage transactions* (typical for a cloud service team) and about the business representation, as we have four applications in our Azure infrastructure, as *how much does application A cost on Azure, considering all the Azure resources involved?*

Question 6: *Which level of detail do you need to reach?* You can choose from the following:

- **High level**: *Is it enough to know that VMs cost X, licenses cost Y, and an internet of things (IoT) hub costs Z?*

- **Detailed**: I need to know the costs of every component—for example, I need to know the costs of the storage, split into the space used and access transactions.

Question 7: *Is the Azure Cost Management tool enough (the filtering, grouping features, drill-down capabilities, and so on) for your purposes?*

Once you have grasped the basic concepts of Azure billing and what information is available with the included tools, you should be able to understand if a different tool is needed for your company.

Question 8: *Do you need to work on the raw data to add business value or maybe integrate into the company dashboarding and a total cost of ownership (TCO) control system?*

This will again depend on how your organization usually maps, displays, and accounts for IT costs. You might already have a billing dashboard, and integrating the Azure information into your own portal or view is the right way of adding cloud spend to your overall picture.

Question 9: *Is the Azure technical terminology right for your audience?*

For example: *Does your audience know what a storage transaction is, or the difference between bandwidth and inputs/outputs per second (IOPS)?*

If you need to export costs to a non-technical audience, you will need to aggregate costs in higher views with custom labels and forget the **Azure Cost Management** tool, or plan to *rename* technical terminology with a custom *dictionary* for non-technical users (for example, general-purpose storage, **LRS (locally redundant storage), and (GRS) (geo-redundant storage)**) into more company-friendly labels such as storage, local storage, and geo-redundant storage.

Having briefly covered how to organize your subscription, and the importance of your hierarchy and initial setup to successful cost governance, one of the first steps to establish full control is to export, analyze, and save your Azure billing information.

Export methods

Keep in mind that this could not be a *one-size-fits-all* approach: you should identify which case is better for you and your organization. And the answer may change over time, prompting you to rework and *evolve* the original approach.

Let's now see the different ways you can get your cost insights.

The Azure Cost Management tool

This is the right tool to start learning about the terminology and how Azure allocates costs and what are the main cost-splitting criteria or cost categories you can find.

The **Azure Cost Management** tool is integrated into your Azure account and it's ready to use.

You can find the tool in Azure by searching for `Cost Management + Billing` in the search bar of the Azure portal (`https://portal.azure.com`).

You can reach the analytics graphs by clicking on **Cost analysis** on the left pane, as illustrated in the following screenshot:

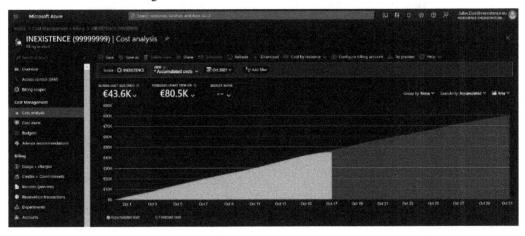

Figure 1.18 – Azure portal: Cost analysis

The default view is a cumulated view of the current month.

> **Important Note:**
> The Azure portal features and views are constantly changing; therefore, you might find your screen a bit different from our screenshots. However, the contents and capabilities are typically only added, so the information will be consistent with ours at the time of writing.

Using the controls in the zone above the graph, you can add filtering clause(s) and change the grouping criteria, the **Granularity** setting, and the rendered graphs, as illustrated in the following screenshot:

Figure 1.19 – Azure portal: Cost analysis (Accumulated option)

In my experience, the most useful filtering and grouping criteria to start with are these:

- **By subscription** (remember that the best practices suggest you split different business services/applications into different subscriptions, as we will see in the next chapter): For example, Production, Disaster Recovery, and so on.

- **Service name**: Simply, the service you bought from Azure. Every service has different `MeterCategory` types.

- **Meter category**: Let's call it `cost-Macrocategory`. It identifies the high-level cost type according to how a specific service is billed (for example, VMs, disks, storage account, IoT hub, **Structured Query Language** (**SQL**) servers, SQL databases, and so on).

- **Meter subcategory**: Useful for a detailed drill-down of cost analysis, it identifies different cost types inside `MeterCategory` (for example, **MeterCategory Disks** has different subcategories such as provisioned IOPS, provisioned bandwidth, storage, and so on).

- **Resource group**: This allows you to filter the cost representation for a *slice*, reflecting the cost of one or more resource groups. If you implemented resource groups *by application*, you could have a first very interesting high-level cost split. Please also note that the cost analysis pane is also available directly on the specific resource group page.

- **Tag**: This allows you to filter based on custom tags applied to the instances. The prerequisite of this view is that you've already implemented a tagging policy, otherwise you may lose instances (and costs!) in the report. We will dig deeper into tagging in the next chapter.

If you are a reseller or provider (for CSP contracts), additional scopes are available to your analysis, such as the following:

- **Billing account** will help visualize billing (pre-tax) for all customers and billing profiles.

- **Billing profile** will help visualize billing (pre-tax) of a billing invoice that can then be filtered by customer or `InvoiceID`.

- **End customer** will display all the costs (pre-tax) associated with a specific customer.

> Tip:
> Additional information can be found here: `https://docs.microsoft.com/en-us/azure/cost-management-billing/costs/get-started-partners`.

In the following screenshot, you will find three pie charts that enable you to have three different views for three additional data dimensions. For example, you may have filtered by **virtual machines** in the main graph and have a different drill-down for **Service name**, **Location**, and **Enrollment account name**:

Figure 1.20 – Azure Cost Management (Cost analysis)

Hint: Azure cost control is a **two-dimensional** (**2D**) tool, and the pie charts add information but can only represent a *static* view and not a trend, so they might not help you in identifying unplanned extra costs if you have very dynamic cloud usage. We will learn how to deal with this later in the book.

To identify what is going on with your Azure infrastructure and costs and represent trends and spikes, you need to represent the costs on a timeline, therefore using the main graph area and switching between different visualizations to obtain a more comprehensive way to represent the same data, as follows:

- Filtering by meter category, grouped by resource group, to identify which resource type is more expensive in which resource group.

- Filtering by resource group, grouped by meter category, to identify, inside the resource group you chose, the resource type that is causing any extra costs.

- You can now filter by resource group, filter by meter category, and group by resource to identify the instance(s) that are causing any extra costs.

You should always consider switching graph filtering/grouping due to the representational limits of the main graph until you identify which resource is causing cost anomalies.

In the following subsections, we'll work on different visualization types for the cost analysis that will get you up to speed with fully understanding the cloud bill and all the related details.

Azure data visualization types

Using the rightmost dropdown in the **Azure Cost Management** main graph, you can change the graph type between the following options:

- **Line** chart: Trend is represented by a line.

- **Area** chart: Trend is represented by a colored area.

- **Column (stacked)**: Each item for the *grouping* criteria is stacked on top of the other.

- **Column (grouped)**: Each item for the *grouping* criteria is presented side by side.

- **Table**

An example of an **Area** chart is provided here:

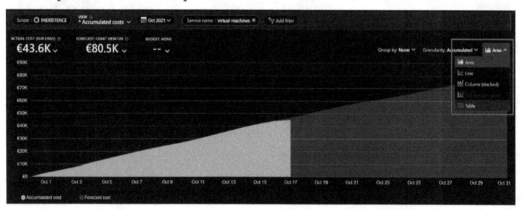

Figure 1.21 – Azure portal: Cost analysis (Accumulated area view)

This visualization is useful to display a trend, and specifically in the **Accumulated** view, since it allows you to identify if the cost is linear (for example, X **Euros (€)** per day, for every day) or if there are spikes or unplanned extra-consumption.

Now, let's look at an example of a **Column (stacked)** bar chart:

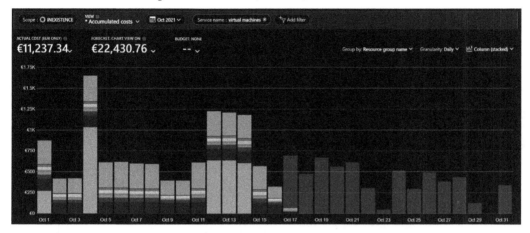

Figure 1.22 – Azure portal: Cost analysis (stacked bar daily view)

This visualization is very useful for period-by-period comparison (for example, with **Granularity: Daily**, to plot every day in the chosen period).

In the previous screenshot, for example, we can identify and compare the costs of VMs for every resource group; it's very easy to identify the spike and the resource group *responsible* for the spike.

Spending forecast

Another very useful feature of **Azure Cost Management** is the analysis of your Azure consumption in the previous period and a calculation of your spending forecast.

The calculated spending for the upcoming future and until the next invoice is typically displayed in a shaded color and will represent your forecasted spending of a future timeframe until the next invoice. Please note you will need at least 10 days of spending data for a forecast to be generated.

Azure and AWS costs

Azure Cost Management allows customers who have a multi-cloud environment to display **Amazon Web Services (AWS)** costs in the same format as Azure spending. This is done via a management group configuration and will allow consistency in billing reports—for example, to your internal departments, regardless of the cloud used.

For more information about how to configure **Azure Cost Management** to import AWS account billing information, please refer to the official documentation at `https://docs.microsoft.com/en-us/azure/cost-management-billing/costs/aws-integration-set-up-configure`.

Amortized versus actual cost

As you open the **Cost Management** tool, the default view is the **ACTUAL COST** view, with all the costs and usage made only in the selected period, as illustrated in the following screenshot:

Figure 1.23 – Azure portal: Cost analysis (ACTUAL COST)

> **Important Note:**
> Please note that the actual cost will still be a few hours behind the actual resource usage, as mentioned earlier.

If you click on **Actual cost,** you are presented with an option of changing it to **Amortized cost**. You can switch to an amortized view by clicking on **Amortized cost** from the dropdown, as illustrated in the following screenshot:

Figure 1.24 – Azure portal: Cost analysis (Amortized cost option)

If you've previously bought upfront reservations, you'll see that the amount will change according to the calculated monthly cost of the upfront reservations.

The following example will clarify this topic.

Let's say you bought, 3 months ago, a VM 1-year pre-paid reservation with a cost of 100 **US Dollars** (**USD**) per month, and an upfront payment of 1,200$. The actual and amortized view will be as follows:

- The actual view will show a spend of 1,200$ the day you bought the reservation.

- The amortized view will show a spend of 100$ each month from the month you bought the reservation up to the chosen timeframe.

We will dig deeper into reservations in the following chapters, but for now, this is only needed to understand the Azure portal's different views.

All costs versus usage only

Another very useful filter when you need to separate the cost representation from reservations (both upfront and monthly) is the **Charge type** filter, which allows you to choose between the following:

- **refund**: If you refunded a previously bought reservation, you can isolate the refund credit with this filter.

- **purchase**: All the purchases you made during the period (for example, licenses, reservations both upfront and monthly, and other services).

- **usage**: The net resource usage, without any kind of reservation purchase(s).

The following screenshot shows the same:

Figure 1.25 – Azure portal: Cost analysis (charge type)

Data export

Every company has its own deck of slides or reports in which you're asked to insert insight and evidence about cloud spending, so you need to export data, tables, or images, and import them into your deck.

There's no need to use the snipping tool: the **Azure Cost Management** page allows you to export data and graphs in **Excel**, **comma-separated values** (**CSV**), and **Portable Network Graphics** (**PNG**) image formats.

You just need to click on the **Download** icon in the upper bar, as indicated in the following screenshot:

Figure 1.26 – Azure portal: Cost analysis (Download data)

An option section will then appear on the right, as illustrated in the following screenshot:

Figure 1.27 – Azure portal: Cost analysis (Download data options)

Here, you can choose the export method (**PNG**, **Excel**, or **CSV**).

Downloading raw data

In the previous method, we learned that the Azure cost control tool is already available for all customers and allows you to understand your costs and the meaning of the various terms such as **Service name**, **Meter category**, **Meter subcategory**, and so on.

This will allow you to explore the costs, but not to integrate this into any of your company's billing dashboards or cost control strategies. Furthermore, Azure cost control does not allow you to *customize* the data visualization, aggregate service names, or rename labels to export a shared view to management or the cost control division.

Fortunately, there is a way, and Azure has made it quite simple for you. When you download your report file in CSV format, you'll see a complete breakdown of all the charges that were invoiced, per used service.

You can export the raw data in a standard .csv file format, open it with your preferred software, and elaborate upon your personal cost report.

For example, you can aggregate VMs, licenses, disks, and NICs in one macro-category called **virtual machines** (in the **Azure Cost Management** tool, you are bound to the predefined categories).

Having all the raw data, it's completely up to you how to aggregate the costs and how to chart them.

> **Important Note:**
> Understanding Azure costs starting from the raw CSV data can be difficult for technical people but practical for a cost controller. Consider practicing using **Azure Cost Management** first, and then double-check your hypothesis on the raw data with **Azure Cost Management** to ensure your understanding is correct and you're exporting the right cost representation.

To export the raw data, go to the **Azure Cost Management** page, select the **Usage + charges** blade in the right menu, and then download the CSV file with the rightmost download icon, as illustrated in the following screenshot:

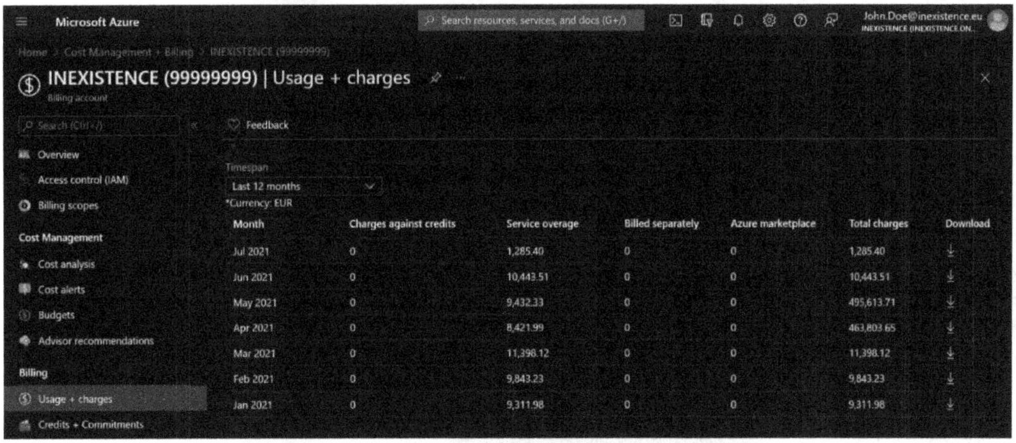

Figure 1.28 – Azure Cost Management: Usage + charges view

If you have an analytics pipeline or data warehouse for cost analysis, you can download CSV files periodically and *load* them into a *cost database* to integrate the Azure costs in your cost control system, and implement your personal representation, aligned with your company standard.

> **Important Note:**
> Viewing and downloading the cost report in the middle of the billing month period allows you to better control costs (since you must optimize your spending on a daily basis, you should not wait until the end of the month) but I strongly advise you to not to count on the month-to-date export to build the monthly cost by adding each delta, since this may not be so reliable.

A daily export is extremely useful for identifying issues and spikes and behaviors, but the best way to load the final stable view for the monthly spend (corresponding to the invoice) is to wait until at least the second—or, better—the third day of the following month (for example, consolidated, stable raw data from January will be available on February 3). This is due to a delay in processing Azure consumption and different time zones. In my personal experience, the third day of the following month is safe.

Automatically exporting daily costs

If you want to integrate all Azure costs in your company cost control pipeline, a manual daily export to *ingest* in your cost control system is not practical: you need to automatically have the costs imported into your system.

Azure helps you with the **Exports** feature, which allows you to schedule an automatic job that exports the CSV file in a storage account. Your cost control system should only download the last CSV file from that storage account and ingest the data automatically.

Just go to the **Exports** blade—here, you can see the already defined export job or create a new job, as illustrated in the following screenshot:

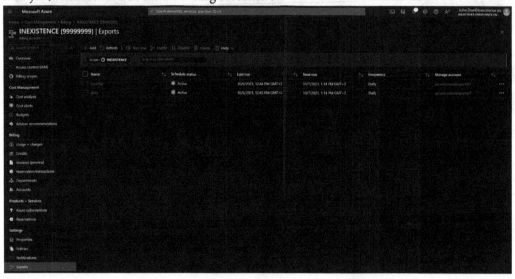

Figure 1.29 – Azure Cost Management (Exports)

To create a new job, you have to click on the **+ Add** button, and then fill out the form displayed, as follows:

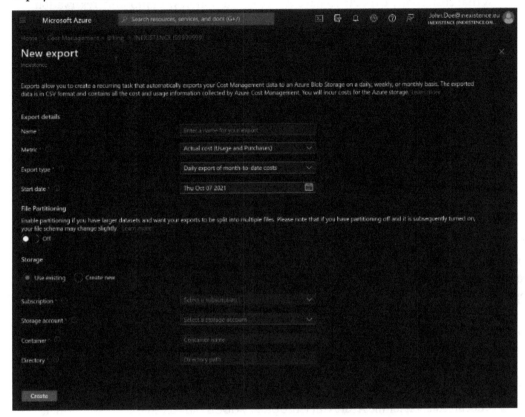

Figure 1.30 – Azure Cost Management (Export options)

Once you named your job, the most important things to set are these:

- **Metric**: Actual costs (without upfront reservations) or amortized costs (with the monthly cost of the upfront reservation)

- **Export type**: Daily export (month-to-date); weekly report (last 7 days), monthly report (last month's costs); one-time report

Other fields let you point the export job to the right storage account, using an existing one or creating a new one, but the most important ones are these two.

Hint: I suggest implementing two exports, as follows:

- **Actual costs; daily export**: To have Azure consumption with monthly reservation and Azure usage costs

- **Amortized costs; daily export**: To have Azure consumption with upfront reservation too, in order to know (by difference) the amount of upfront reservation monthly

> **Important Note:**
>
> With your tagging strategy if you need to represent costs by custom tags, you'll find a field named **Tags** in the exported CSV file, with a **JavaScript Object Notation (JSON)** data structure with all the tags (the ones you added and other service tags, normally hidden).

Using the command-line interface

Another way to get Azure detailed costs is the **command-line interface (CLI)** (both the PowerShell and the Linux AZ CLI). This is a very interesting option to easily generate reports that you can send to key people, directly with simple batch files, without having to struggle with exporting from Azure, importing the CSV file into a database, and configuring a **business intelligence (BI)** tool.

> **Tip:**
>
> To install the previously mentioned tools, please refer to the official documentation for the Azure CLI (https://docs.microsoft.com/en-us/cli/azure/install-azure-cli) and PowerShell (https://docs.microsoft.com/en-us/powershell/azure/install-az-ps?view=azps-6.5.0).

Please note that using the Azure CLI (fired within the web browser, from the portal) will incur costs derived from the storage you use and the data transfers, for it to work. Using the Azure Cloud Shell will always have the latest tools available for Azure CLI and PowerShell. Since you need to login through the browser to use the cloud shell, your session is always authenticated. There is a very minimal charge for the static data stored in directories in cloud shell, which uses a storage account in Azure.

The documentation starting points are listed here:

- **AZ CLI—Billing** (https://docs.microsoft.com/en-us/cli/azure/costmanagement): Lets you manage invoices, billing profiles and details, configure roles for users, list products for billing account, and so on

- **AZ CLI—Consumption** (`https://docs.microsoft.com/en-us/cli/azure/consumption`): Lets you manage usage details, reservations, and budgets, and view price sheets

- **AZ CLI—Cost management** (`https://docs.microsoft.com/en-us/cli/azure/costmanagement`): Lets you manage **Azure Cost Management** exports (refer to the *Automatically export daily cost* section)

- **PowerShell** (`https://docs.microsoft.com/en-us/powershell/module/az.billing`)

Though the official documentation may point out a lot of different parameters and queries for managing the cost control by command line, I will focus your attention on a couple of commands for extracting and managing the usage details.

Let's focus on the `consumption` subcommands, which allow us to download information on currently generated costs on our account.

Here are some AZ CLI examples.

To get Azure usage details for a specific period (a day or a month in the past), you can use the following command:

```
az consumption usage list --include-meter-details --include-
additional-properties --start-date 2020-10-01
--end-date 2020-10-02
```

This command will return JSON with useful information (among other useful information) such as the location, instance ID, subscription, resource group, and net costs (without taxes). This command is supported only with EA, Web Direct, and MCA offer types.

> **Important Note:**
> If you prefer to have the output in a **tab-separated values** (**TSV**) format, you may add `--output tsv` in the command line, but you'll lose any nested information. Please try executing the command for the JSON and TSV formats and compare the output: you'll find that the JSON format is far more detailed than the other one.

Important things to keep in mind are listed here:

- The command line allows you to have the *net* prices.

- The command line allows you to specify a period start/end date.

- The command line allows you to get the *top N* spending resources.

So, it's very useful to have a quick overview of the costs and to work directly in batch scripting, but it's not comprehensive and it does not represent the final invoice.

Please note that on the tagging strategy, you'll find a JSON element with all the instance tags. This information is lost in the TSV format.

Using application programming interfaces

If you want to better control data ingestion, instead of downloading a CSV and ingesting it in your cost control system, you can do it programmatically by using **application programming interfaces** (**APIs**). Microsoft has different APIs that let you download the billing details, forecast, budget, and so on.

The main documentation entry point is `https://docs.microsoft.com/en-us/rest/api/consumption/`: this is where you should start reading documentation about **REpresentational State Transfer** (**REST**) entry point(s) you need to invoke to get useful information about your billing, such as the following:

- Usage details—the full CSV or JSON structure for the billing of the desired period (this allows you to customize the period start and end dates)
- Reservation details
- Pricelist
- Budgets
- Forecasts (only for Enterprise customers)

In addition, at the `https://docs.microsoft.com/en-us/rest/api/billing/` link, you will find information about other operations on billing objects (for example, transfer billing ownership, payment methods, and so on). Usually, if you're searching for a way to control your costs and integrate the costs in your pipeline, this is not the first point from where you should start.

> **Important Note:**
> If you searched online, you may find old Reporting APIs (`https://docs.microsoft.com/it-it/rest/api/billing/enterprise/billing-enterprise-api-usage-detail`). Despite being officially discontinued, they still work, but we recommend avoiding these APIs for building an injection.

As you can see from the official documentation, using APIs allows you to deeply integrate billing data into your dashboard and cost management software, but it may be a little complex at the beginning, considering you don't know what to expect from each API and you'll probably end up searching in different exports for missing billing details you're desperately seeking.

Question 10: *But APIs are a bit complex—why are they so useful?*

Unfortunately, there is no simple answer, but APIs are generally very smart for integrating analysis, feedback, and decisions based on an external tool. Just think about an enterprise that uses an integrated authorization flow for any *action* that could generate costs: creating a VM will generate costs, adding a disk generates costs, creating a new SQL database or an Azure Kubernetes cluster generates costs, and all should be authorized internally.

Your company probably has a centralized governance team that can provide users with internal processes that automate the provisioning of cloud resources. So, everyone should compile some form in a software tool to ask and wait for approval, which will then trigger deployment. The tool will use APIs to start the resource creation process in an automated fashion.

The same applies to buying reserved instances or to cost accounting inside the enterprise: you might need to extract costs based on the department or management group.

> **Important Note:**
> You cannot perform this kind of operation without using APIs: it's the only way to integrate Azure actions and extractions, programmatically, to an already existing software (generally an enterprise-scale software).

In the end, please keep in mind that Microsoft itself suggests you use the method we discussed in the *Automatically exporting daily costs* section for large data handling in its official documentation, which can be found at `https://docs.microsoft.com/en-us/azure/cost-management-billing/costs/manage-automation`.

In conclusion, my personal suggestion is to start from the first method—the **Azure Cost Management** online web page—to better understand the terminology and how costs are grouped and exported by Azure.

The next step will be downloading a full monthly invoice, opening it with your preferred spreadsheet, and studying each row, eventually pivoting and comparing what you find in the CSV file and what you've found with the **Azure Cost Management** tool. Then, you can import the CSV file into a database (a simple open source database is more than enough) and start querying the database, grouping results, and constructing tables to *translate* the technical terms into something more understandable for non-technical people (such as cost controllers or C-levels).

One more step in automating cost control ingestion is to schedule one or more exports and download the file(s) from the target storage account.

The last, more complex but extremely flexible way to integrate cost management and implement a feedback loop is to use the **Consumption** and **Billing** APIs. Once you have your fully automated database ingestion, custom tables, and whatever you need to customize cost representation, you can build your own BI dashboard on it and master cost control.

Summary

In this chapter, we have worked on how Azure bills consumption on several different services in order to understand the invoice at the end of the month. We started with defining what the possible billing types are, then proceeded to the Azure pricing online page, which allows you to foresee and understand how each service is billed and charged. We learned how to price a few key services: VMs, App Service, Cosmos DB, and Azure Functions, which all have very different policies and frequencies of billing.

We then moved to the Azure portal **Cost Management** section and provided a few useful views that can give you relevant information, and learned how to export the preferred views in the possible formats. We considered API options as a source of billing information, although this book will be focused primarily on the information available through the Azure portal.

In the next chapter, we will focus on how you intend to manage cloud costs and how this can be configured and automated with the available Azure tools.

Questions

1. What is your billing type?
2. How can you provide a cost analysis with a daily stacked bar chart view of the last quarter?
3. How can you export your billing information?

Further reading

- Management groups and hierarchy: `https://docs.microsoft.com/en-us/azure/cloud-adoption-framework/ready/azure-setup-guide/organize-resources?tabs=AzureManagementGroupsAndHierarchy`

- Well-Architected Framework: `https://docs.microsoft.com/en-us/azure/architecture/framework/`

- Cloud Adoption Framework: `https://docs.microsoft.com/en-us/azure/cloud-adoption-framework/`

- Azure cost analysis: `https://docs.microsoft.com/en-us/azure/cost-management-billing/costs/quick-acm-cost-analysis`

2
What Does Your Cloud Spending Look Like?

In the previous chapter, we laid out the basics of understanding your Azure cloud costs, how to use the pricing tool to understand each Meter Category, and how to read the cost analysis and export from the portal all the information relevant to your company. It is now time to associate the spending of your cloud services with the business stakes of your company and start thinking about how to reduce that spending to the minimum possible while guaranteeing the desired service levels, which is the equivalent of a cost optimization mindset.

We'll learn how to deal with the following:

- Defining cost constraints
- Aiming for scalable costs
- Paying for consumption
- Reviewing the subscription hierarchy and management groups
- Understanding cost optimization automation and policies

Upon completion of this chapter, you will have all the tools and guidelines to successfully aim for structured cost governance that matches the financial and technical **key performance indicators** (**KPIs**) of your company.

Technical requirements

To follow the instructions in this chapter, you'll need the following:

- A computer with internet connectivity

- Access to the Azure portal, with a working subscription and the related credentials

- Privileges to access **Cost Management** information (see `https://docs.microsoft.com/en-us/azure/cost-management-billing/costs/assign-access-acm-data`)

Defining cost constraints

We now need to apply what we have learned to our company's organization, for costs to be clear and split across all departments and for technical constraints to be correctly implemented and configured for our cost governance model.

Depending on your cloud usage and how your organization works, you must create a conceptual map on how a chargeback (albeit theoretical) should work for each cloud service. If your organization is made up of many separate departments, each holding its own authority over cloud costs, these suggestions will still work, although you might have to duplicate all your efforts.

For example, in a company where a central **information technology** (**IT**) department is providing services to other departments, these could be the criteria:

- **Application usage**:

  - Each application has a unique and dedicated set of cloud services that can be charged back to users (one-to-one relationship app/department/cost), as illustrated in the following screenshot:

Figure 2.1 – One department has one application with its costs

- Each application is shared between the departments and chargeback is equally distributed (one-to-many relationship app/department/cost), as illustrated in the following screenshot:

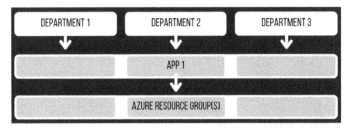

Figure 2.2 – Three departments share one application, splitting costs evenly

- Each application is shared between the departments, and chargeback has a different weight according to each department (one-to-many relationship app/department/cost), as illustrated in the following screenshot:

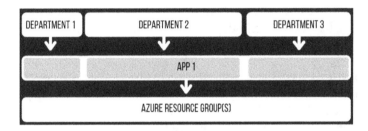

Figure 2.3 – Three departments share one application, splitting costs proportionally

- **Shared services where applicable** (such as networking, storage, governance, automation, and so on):

- These can be bundled together and charged to the common IT department.

- Or, they can be split between the application owners according to the application usage rules we defined.

> **Tip:**
> Networking services are a family of charges that typically are very difficult to calculate for chargeback—we will learn how to do this in the upcoming chapters.

- **Platform-as-a-service (PaaS) services and development-operations (DevOps) deployments**:

  - These costs are typically quite tough to split, and the best practice is to try to keep the applications in this realm isolated as much as possible, in order to contain all costs within the related application.

> **Important Note:**
> The key takeaway here is that you should try as much as possible to reflect the business owners' KPIs into this map so that it will be easy to put a price tag on the current, past, and future spending for a specific owner.

Once you have built a cost chargeback map of your business needs, this must be shared with the IT department and cloud architects so that they can translate the technical requirements for this map to be in place.

Let's work on a practical example to explain the principles. Our sample company, Inexistence.eu, has several main departments, each with its fair use of cloud services, as follows:

- **IT**: All the main stakes of the IT department are now in the cloud: from provisioning apps to security, internal ticketing, and so on. Since there are many applications that are old and under obsolete platforms and **operating systems (OSes)** and languages, these are mostly **infrastructure-as-a-service (IaaS) virtual machines (VMs)** with fast storage options and large databases that require critical support.

- **Marketing**: The marketing people run campaigns through social media and the web to attract people to the website; these applications are all PaaS services due to the nature of the campaigns, which must be very fast and with little maintenance.

- **Sales**: Sales tools are a critical aspect of the company, and these are 24x7 hybrid architectures with IaaS and PaaS and rely on the shared services.

- **Operations**: The operations department deals with issues such as predictability and processes that are governed neatly; the applications used are all for monitoring and addressing issues.

- **Finance**: These guys mostly use Excel files and typically like to extract data to re-elaborate. They typically need clarity of information and precision, and their applications are mission-critical and usually compliance-constrained.

- **Customer care**: This is the frontline of your company's interactions with your own customers. Any lag in these applications will result in poor customer service. The applications and integrations are mission-critical and performance-bound; every second counts when there is an issue with customers.

As you can see, simply dividing the costs by application is a very complex exercise, especially when apps are even shared among departments and the usage weight is unknown.

One option is to define a subscription for each department and a common subscription for all the shared services: each department will have a chargeback of its subscription(s) costs, plus a subset of the shared services that can be defined by calculating the spending quota of each department.

If you are starting from a greenfield, then the **Cloud Adoption Framework** will help you tackle these decisions (`https://docs.microsoft.com/en-us/azure/cloud-adoption-framework/decision-guides/subscriptions/`) and configure the right hierarchy. However, many customers find themselves starting from mixed configurations and will need to find a way of defining cost constraints that work in their complex environments.

Here's an example. Consider the following subscriptions for different departments:

- Marketing spending is 100,000 **US Dollars (USD)** $/month across all its subscriptions.
- Sales spending is 80,000 $/month across all its subscriptions.
- Customer care spending is 120,000 $/month across all its subscriptions.
- Shared services spending is 100,000 $/month across all its subscriptions.
- Total spending: 400,000 $/month

Here is the quota of shared services:

- 34% marketing, which will be billed at 100,000 $/month + 34,000 $/month
- 26% sales, which will be billed at 80,000 $/month + 26,000 $/month
- 40% customer care, which will be billed at 120,000 $/month + 40,000 $/month

This is just an example of how you can split shared services. Different quota criteria can be applied and strictly depend on how your company will value charged services.

Unfortunately, very often, cost control cannot be completely split by infrastructure design. In reality, during your cost isolation planning, you must consider the following:

- One object (technical resource) can exist in only one subscription.

- **Virtual networks** (**VNets**) cannot, at the time of writing, interconnect objects in two (or more) different subscriptions, without creating one or more VNet peering(s), thus adding extra costs for network traffic.

- Rescaling IaaS and PaaS services might require, depending on the service and its configuration, (albeit minimum) downtime (so if they're shared, you'll end up having multiple business application downtimes).

With these considerations in mind, you may start thinking carefully about what exactly the shared services are now, and what is your cloud strategy for *tomorrow*.

Let's see a few examples now.

We are deploying a new PaaS service for the **Marketing** subscription.

Let's look at the next two questions.

Question 1: *Will this service be completely dedicated (today and tomorrow) only by the marketing division?*

Question 2: *If this service is in your cloud adoption roadmap (or strategy), should it still be considered dedicated now to become shared tomorrow?*

If you consider it as a shared service (such as **Azure Active Directory** (**Azure AD**), Azure ExpressRoute, Azure Application Gateway, Azure Firewall, Azure Monitor, and so on, and in general any workload that can be shared across many departments at an application level) from the beginning, any other new project will find it already available, cutting down the time and cost of recreating and reconfiguring existing applications and integrations.

We need to set up the backup of a PaaS service.

Question 3: *How do you consider the backup of PaaS services?*

Normally, any PaaS backup is stored inside the same subscription of the service, so it's already included in the *per-subscription billing* report, and you could isolate that cost.

Question 4: *How do you consider the backup of IaaS components by Azure Backup (a PaaS service)?*

Normally, you should define one or more backup vault(s) for any subscription that has VMs to be backed up.

Question 5: *How do you consider the backup of IaaS components not supported by Azure Backup?*

(This question is deliberately challenging since you need to know in advance which services installed on each VM are compatible or not with Azure Backup: some databases, for example, are not supported; neither is disk striping.)

If you need to have a separate backup infrastructure, you need to carefully plan the backup traffic, as follows:

- Having the backup server(s) in the *shared* subscription may amount to extra costs due to the backup traffic over all the peering connections.

- Depending on the regional architecture and company governance model, the network traffic costs might still be cheaper than, for example, duplicating servers for each region.

- Having one backup server in every subscription that has third-party backup services may bring you licensing and IaaS extra costs, and more operations to manage every different server, storage resource, and so on. In addition, having more backup servers will negatively affect compression and deduplication.

So, based on these few simple examples, it's clear that a cost-splitting policy should be built on a clear, defined, and very solid cloud adoption strategy that considers all the technical aspects of your virtual public cloud data center.

We've learned in this section how to split costs according to your company's organization. Several types of configuration and automation tools allow us to get closer to the overall cost by application picture, and you might need to integrate parts of your internal systems to adapt to the cloud operating costs. We will dig deeper into these tools in the last section of this chapter; meanwhile, the key takeaway is *you need to know who owns which cost in your organization*, even if the costs are not really charged back.

If there aren't any cost owners, you should request they be created. Ownership of cost is essential to healthy cost governance since it will allow the establishment of a virtuous process of improvement against measurement for right-sizing and cleanup. Most organizations where there is no clear application owner are failing to address a cost governance motion and will soon realize once they have an owner that this will be much easier. In addition to the cost ownership mentality, lots of technical configurations and implementations will be added to complete a full cost governance process, as we'll learn throughout the chapters of this book.

Let's move on to the next section and see how costs should be scalable, to be able to move the billing needle according to usage, not allocation.

Aiming for scalable costs

Let's now learn how to organize the application infrastructure to be able to scale resources up and down, matching the real usage and not the original provisioning sizing.

In legacy IT infrastructure, **scalability** is typically defined as the ability of a program or application to keep running and working healthily when conditions change in traffic volumes and/or performances. Scalability can be divided up as follows:

- **Horizontal scaling**: The program or application will add more compute instances to its architecture to make up for the increased volumes.

- **Vertical scaling**: The currently allocated servers will have an increase in terms of the **CPU, RAM,** or storage.

The benefits of scalable applications are clear, especially within cloud infrastructures: by increasing and subsequently decreasing the capacity according to peak demand, not only can we address performance issues, but the application can also be hardened and can be reduced to a minimum set of services at night or at low traffic moments.

In the cloud, where its flexibility and pay-as-you-go policies have been the most coveted features, scaling has a direct impact on costs: if you need to scale your application to thousands of servers, the cloud has your back, but be prepared to pay the price at the end of the month. On the other hand, you should never be shy and let a scalable service run at peak when it can be downsized: another paradigm of cloud cost governance is that *every little counts*, and spending that may seem trivial at a first glance can reveal itself to be daunting at the end of the month when the service sums up all the consumption.

In your company's virtual data center, you will have three main types of applications/ services in an application, outlined as follows:

- Those that are not ready to scale in both a horizontal and vertical fashion

- Those that are scalable by adding servers/instances (horizontal)

- Those that are scalable by adding resources (vertical)

The architecture of each application thus has a direct connection with its costs, and several seemingly small changes to it can lead to higher costs or savings.

Scaling applications—when supported—is an exercise of the following:

- Analyzing the application architecture to identify how it could be scaled (services, protocols, application servers, databases, batch or scheduled jobs, load balancing algorithm, and so on)

- Performance optimization

- On-demand or on/off patterns' application

- Scaling down, releasing unused resources

Scaling applications—when not supported—is an exercise of the following:

- Analyzing the application architecture (services, protocols, application servers, databases, batch or scheduled jobs, load balancing algorithm, and so on)

- Defining if and how the cloud infrastructure may allow you to *trick* the application and scale it anyway

For example, let's consider an application provisioned by a template for which, for any reason, you no longer have a maintainer/developer/supplier; the application receives some documents, then saves the files in a local folder and the metadata in a database. It is composed of a single VM, sized for 100 concurrent users, but you only have 90 users in the first 5 days of the month, and an average of 30 to 60 users during the rest of the month.

You need to allow it to scale, and therefore analyze each component on the full-stack VM—every configuration and endpoint. Once you know the application's architecture, you have to decide whether to maintain it as-is because the risk of a fault is too high, or split the application by editing configurations and endpoints manually.

In the following diagram, you can see a full-stack application on the left and a split component on the right, with a *dotted* representation of horizontal-scale components, behind a load balancer with source **IP** port affinity:

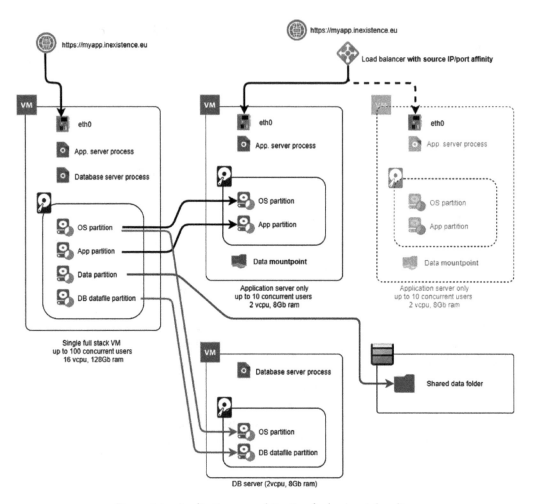

Figure 2.4 – Application re-architecting for horizontal scaling

Therefore, with a little configuration tweaking, you will be able to individually scale the frontend components as if the application were to support a scaling method.

Serverless resources from a cost perspective are the most optimized both in terms of costs and general resource consumption: they are spun up and charged for only when you use them, and once the execution is over, they immediately stop consuming.

Even when you have a monolith application with a single server, you still have the option to scale it up and down vertically—for example, by using Premium disks during peak hours and scaling them down to a lower tier drive as Standard HDD (for Ultra disks, this could be lowering **input/output operations per second** (**IOPS**) and throughput to the minimum, or even hot-swapping them) at night when VMs are switched off. Or, you could use a cheaper VM option during weekends but resize it on Monday mornings to get the full capacity of a more performing VM or offload data that is not frequently used for colder and cheaper archives.

Another important part of governing cloud costs is to be able to enforce policies for applications that can be handled in the following ways:

- **Switched on only when needed (on-demand cost governance)**: This is typically useful for development and non-production workloads that are not used often and can therefore be switched on only when needed. This can be automated and even offered via a user portal or other easy means for application users to be able to access the infrastructure when needed.

- **Automatically switched on and off (scheduled on/off cost governance)**: This is normally used for non-production workloads that can be off at night and at weekends when nobody is using them. Rather than offering on-demand, this is useful for complex workloads that have many co-dependencies when booting, hence the whole application chain can be scripted and automated to correctly shut down and turn on at regular intervals. This makes for consistent cost governance and can allow many saving options, as we will learn in the following chapters.

- **Automatically scaled on/off according to usage (demand shaping scalable cost governance)**: This is the most sophisticated form of cost governance of an application, and typically the most efficient. Rather than simply scheduling on a flat static timeline, this method allows for a script to dynamically identify the resource usage of an application and respond quickly by scaling it up and down. It is particularly efficient for PaaS products, where you can shape their tiers (and costs) easily without caring for infrastructure details. Obviously, and depending on the resource, you need to make sure that the final state (deallocated, and so on) is one that is not billed.

> Important Note:
> The key takeaway here is that you need to start thinking of your applications' and resources' infrastructure as something *fluid* that will always change over time and can be sized up and down to meet your performance and cost needs.

Most customers start this difficult conversation with the sentence, *my applications are not scalable*, or *I cannot switch off production workloads*, and end up, a few months later, with every possible scaling of each and every application to get the most out of the cost governance process.

Here are a few questions on finding the right way to scale a *legacy* or *standard* application:

Question 6: *Are all the resources (of every environment) used constantly up to 60% to 80% of the capacity?*

If not, you can immediately perform downsizing, finding a new resource tier (**SKU**) that matches your current usage, to have some quick savings.

> **Important Note:**
> Dynamically scaling resources is a native-cloud paradigm, but unfortunately, not all applications will support it or allow it. Before downsizing or making any cost-related change, you should always check with the application owner if the requirement for those resources is tied to a traffic peak for which the application cannot scale, or if it is a 24/7 application that can never be switched off.

Question 7: *Does the application have one or more non-production environment(s)?*

If yes, they're the right candidate to be switched off on a scheduled basis or converted to an *on-demand* environment.

Question 8: *Is your (web) application stateless?*

A stateless application is perfect to be horizontally scaled since you can power up new nodes when needed, shut them down when no longer needed, and even take advantage of spot VM instances at a lower price. A stateless application is also a perfect candidate for containerization, and you can also use Ephemeral OS disks, which are free.

In cases where the application is stateful, you could horizontally scale it only if the session is shared between all the application servers or is maintained outside the horizontal-scale VM pool.

For example, in the **JBoss** open source alternative **WildFly**, at the time of writing, you could configure a shared session between different application servers (allowing for horizontal scaling), adding proper values for a `shared-session-config` node in the `configuration.xml` file.

If you want to decentralize the session content, you may also rely on an in-memory distributed data store/data grid.

> **Important Note:**
> Please consider that on application startup on a new node, it may require some time to announce the new instance to the grid.

Question 9: *Is your (web) application stateful and does it use shared user sessions between application server nodes?*

This application is perfect to be horizontally or vertically scaled: since the user sessions are shared between the nodes, your users will not lose their work if you shut down and resize a resource to a different tier, at any given time.

Question 10: *Does your application perform compute- or memory-intensive operations during specific moments (for example, on the first day of the month)?*

This application is suitable for vertical scaling because you can shut it down and change the tier of the resources to bigger and better performing resources for only the peak day, and then go back to a smaller tier for the remaining period.

> **Important Note:**
> Before doing any downsizing or changes in production, always check for resources' quotas and availability in the selected region/subscription.

In addition to scaling from the application, there are several objects in Azure that will help keep down costs by providing scalable compute resources, such as Spot VMs or B-size VMs, VM Scale Sets, or Azure Batch. We encourage you to work with your technical team to find the best approach to the choice of infrastructure.

We know that scalability is not just a pretty name: it is essential to the health and performance of any application and is also vital to cost control in a public cloud environment. We will dig deeper into this in the next section.

Paying for consumption

One of the main obstacles to great Azure cost governance—and, to be honest, to any cloud cost governance—is the *old* IT misconception that a VM is something known, beloved, and cherished, much like a pet; while in the cloud, a VM is just a service, a commodity, and should be switched off and on according to its real usage.

According to the post at `http://cloudscaling.com/blog/cloud-computing/the-history-of-pets-vs-cattle/`, servers that are given a friendly name and are considered indispensable and unique objects with critical support are what we call **pets**; they are manually operated and their owners have a clear affection for every single one of them. If they fail, everyone knows they are down, and the IT folks must immediately work on restoring them until they are back and shiny.

On the other side, arrays and systems built via automation, such as clusters, racks, and anything that is load-balanced, are what we call **cattle**; they can be switched off without a significant impact on the applications. If they fail, others will be quickly brought up, replacing the faulty ones. The old thinking of *redundancy* is also a bit old for the cloud: there can never be too many instances of any application that is designed to properly scale, and considering failure as part of the application helps with this. Architecture has a clear saying in defining when a server is a pet or cattle: even with redundant systems of large databases or file servers, a bunch of servers does not constitute cattle but more a *pet set* – just a larger number of servers that are manually operated and cannot fail.

What has changed with the cloud is this:

- We can spin up VMs at will and with great speed
- We can resize and change **hardware** (**HW**) requirements in a couple of clicks
- Scalability is no longer an issue: assume you can switch on as many VMs as you like
- VMs are disposable
- Containers are the perfect cattle example
- **Continuous integration/continuous deployment** (**CI/CD**) and **infrastructure as code** (**IaC**) have expanded the cattle concept

Therefore, it is imperative that you, as the cost governor, establish a culture of pay-as-you-go, and pay as necessary, trying as much as possible to fight the old urge of adding memory and resources *as they might be useful later on*. You are paying for resources now, and if you don't use them, you must turn them off.

The key concept of the cloud is that you are paying for pure consumption and should never be afraid or sorry to switch a service off, especially when it can be restarted in minutes. And yet, many large enterprises and companies struggle to grasp this concept and often prefer to reserve a staggering amount of wrongly sized VMs rather than admitting that they are pets for them and switching off the unnecessary consumption, finally realizing that they are cattle.

Once you have established which are your pets and cattle in your virtual data center, you will also know where you will be able to splurge and where you can make significant savings in your cost governance process.

You must be ready to know in your application map which applications are fault-tolerant and embrace failure so that you will be able to *play* with right-sizing and down-sizing without fear of service interruption.

Question 11: *Have you identified your pets? And your cattle?*

Let's see an example of this in practice.

You have an application running on IaaS infrastructure. The Azure region is North Europe. It is composed of the following:

- One backend VM

- Two frontend VMs with a load balancer

You check with the resource usage and find out the following information:

- **Backend VM**:

  - Size: D4 v2; 8 **virtual cores** (**vcores**); 28 **GB** RAM; 400 GB temporary storage. The cost of this resource is around 648 **Euros**/month. Its CPU usage is below 5%; memory usage is peak 16 GB.

- **Frontend VMs**:

  - Size: 2x D4 v2; 8 vcores; 28 GB RAM, 400 GB temporary storage. The cost of these resources is around 1,300 €/month. Their CPU usage is below 10%; memory usage is peak 16 GB.

We'll see the details of cost savings techniques in the next chapters, but at this point, you need to know that the main steps to cost governance for this example are the following:

- **Downsizing**: Define a cheaper, more tailored model of VM both for the frontend and backend.

- **Switch off/on**: Talk to the application team and see if the following can be done:

  - The whole stack can be switched off at night and at weekends—the cost savings of running can be up to 50% of the original cost.

- If not, consider keeping a lower profile and do the following:

  - Only use one frontend VM for off-peak operations—this will reduce the spending up to 30%.

For the same VM, you found out you are using all Premium managed disks, which is basically a PaaS service. You decide to make the following changes:

- Switch all the OS disks to Standard **SSD disk** or Standard HDD (if the application does not reside on the OS disk).

- For the frontend VM that is switched off during off-peak hours, you prepare a script that will bring the storage tier to Standard HDD, then when you switch it on again you move it back to Premium. This way, you are not paying premium storage for a disk that is not even used.

As we saw in the examples, we need to practice being extremely precise in terms of resource usage and especially during the time they are used: in an on-premises data center, keeping all the servers on is not a big deal (except maybe for the energy consumption), but in the public cloud, you need to run only the resources you are using and for the time and tier they are needed, and this is valid for IaaS and PaaS services, as we'll discover in the next chapters.

In this section, we analyzed how to deal with consumption services. Let's now dig a bit deeper into how your subscription hierarchy can help you govern its costs.

Reviewing the subscription hierarchy and management groups

We learned in the previous sections how to split and associate costs according to your organization's needs. In this section, we will learn how to define a subscription hierarchy that matches as much as possible your company's finance operations, and we'll also introduce technical concepts that are relevant to the cloud optimization process, such as management groups.

As a rule of thumb, I have seen customers make the following choices over time, each of them changing and reverting to the other where possible, mostly due to the limits in charging each service:

- **Keep all resources in the same subscription**: It seems logical in the concept of a single virtual data center to keep billing all in one large bucket. Please keep in mind, though, that using one single subscription for all your resources has several limitations: in terms of technical resources being limited, and in terms of chargeback operations being more difficult. The current best practice, where possible, is to define several subscriptions, each with its **segregation of duties (SoD)**.

- **Define one subscription** per department/entity/chargeback account/application or set of applications.

- **The DevOps team**: This deserves a section of its own, for the implications of letting developers spin up their resources on costs can become unpredictable without control, and yet too much control will render the DevOps team paralyzed. Finding the right balance that will allow you to control costs while delegating innovation and operations to the dev team is key to a successful DevOps process.

- **By tags**: This approach is related to cost control and reporting only and will be outspread in a dedicated section later.

Management groups

When customers started implementing as a best practice segregation of services into several subscriptions, the main problem became how to govern and address many subscriptions. Hence, **management groups** were created to allow customers to define an aggregated view of one or more subscriptions and, most of all, to be able to apply policies and privileges at an aggregated level, without having to apply all the changes to every single subscription.

A subscription hierarchy can now be configured and, most of all, governed via policies (which we'll see later in this section), as displayed in the following screenshot:

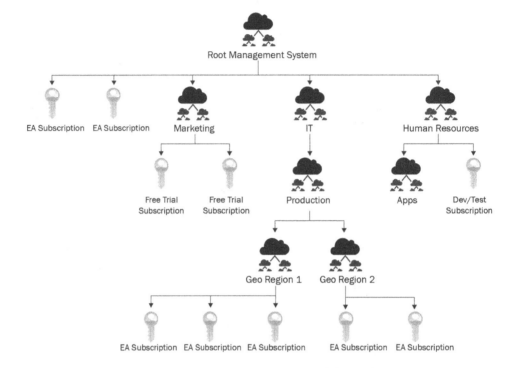

Figure 2.5 – Management group sample hierarchy

Basically, management groups provide a logical level above subscriptions, which can be used to apply policies and governance to one, a set, of or all your company's subscriptions. A few important highlights of management groups are provided here:

- Each management group is linked to the same Azure AD tenant, and each directory has one management group hierarchy.

- A single directory can support up to 10,000 management groups, across six levels of depth (root level excluded).

- A management group can only have one parent but can have many children.

- The top-level management group is present by default and is called **root**, and includes all of your subscriptions until you create a new hierarchy. The root level is used for policies at global and directory levels.

- All activity for management groups is logged in the **Azure Activity Log** page.

If you haven't used management groups already, let's dive into it and create a simple hierarchy, as follows:

1. Navigate in the Azure portal to the **Management groups** pane, as displayed in the next screenshot:

Figure 2.6 – The Management groups initial pane

2. At first use, **AD Global Admin** access is required to elevate to the **User Access Admin** role of the root group. After this first step, the admin can assign other users or groups to manage the hierarchy.

3. Click on **+ Create**. The management group pane appears, as shown:

Figure 2.7 – Creating a management group

4. Choose a name for the management group that reflects the hierarchy of cost for your organization. In our example, we might choose **marketing** and use this group for all subscriptions that are dedicated to the marketing department and initiatives.

5. Then, click on **Submit**, and the management group is created right after the root group, as illustrated in the following screenshot:

Figure 2.8 – The newly created management group

6. The newly created marketing management group is empty. You will have to choose which subscription will be moved under this management group.

    If you want to move subscriptions to the new management group, navigate to the management group display window, click on **Subscriptions**, and click on **+ Add**, as shown in the following screenshot:

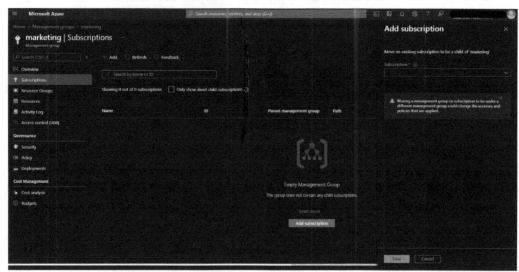

Figure 2.9 – Moving subscriptions under management groups

In this section, we have learned how to use management groups to reflect the hierarchy of costs that your company will be using for cloud cost governance. This will also affect other aspects of the Well-Architected Framework, such as security and landing zone, but for this book, we'll only be focusing on the cost governance portion of it. The next step will be automating as much as possible every control of the resources by leveraging this management group hierarchy.

Understanding cost optimization automation and policies

Before we address how to control and monitor costs, we need to know the details of how our costs are distributed across business applications.

In large enterprises, it's pretty normal to have a few core applications that spend about 20% more than others and waste a little bit of compute or memory resources: core applications need to be instantly available, thus we need more contingency compared to the non-critical ones, despite every possible horizontal scaling architecture you can realize.

So, the first thing to know when you migrate an application from on-premises to a cloud infrastructure is how the applications work, their flow, dependencies, and so on, and identify all the resources involved in that application, to better concentrate at a later stage on the component's optimization.

> **Important Note:**
> If you don't know how the application is composed and how it works, it's quite unlikely you will reach a good level of cost optimization, although if you start working on how to optimize its costs, you'll likely end up with a good level of understanding of how the application works.

Tagging

To better understand the **tagging** concept related to your cloud environment, please forget about infrastructure complexity, the organization teams, units, and processes, and let's focus only on the concept of describing a shipment box in a warehouse full of boxes, as shown:

Figure 2.10 – How to deal with lots of indistinguishable boxes

What do the shipment companies do? They attach a label with text, a barcode, or a **Quick Response** (**QR**) code that describes every single box.

You can do the same thing with cloud resources: you can add a **tag** to the resource, to better describe its scope and features.

> Tip:
> Tags are key/value pairs, but this does not limit you to assigning more values to a tag if you define your own standard (for example, a field separator character).

How to tag a resource on Azure

As we just mentioned, tags are key/value *string* pairs that you could set via the following:

- **Azure portal**: Manually, using the **Tags** blade as shown:

Figure 2.11 – An example of several tags on an Azure managed disk

If you need to update tags on more resources, you could do this in one single operation: you have to research them using the correct criteria, select the resources you want to manage the tags, and then use the **Assign tags** button over the filtering section, as illustrated in the following screenshot:

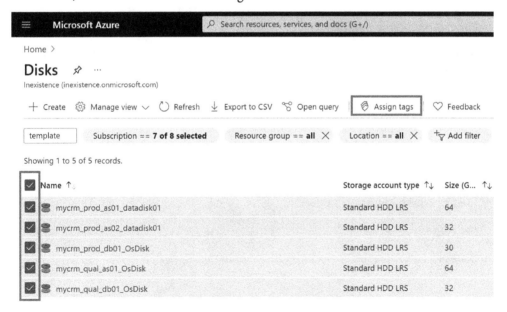

Figure 2.12 – An example of bulk-adding tags

- **Azure command-line interface (CLI)**: With the `az tag` command, as described in the official documentation (`https://docs.microsoft.com/en-us/cli/azure/tag?view=azure-cli-latest`)

- **Azure PowerShell cmdlets**:

  - `Get-AzTag` (`https://docs.microsoft.com/en-us/powershell/module/az.resources/get-aztag`)

  - `Remove-AzTag` (`https://docs.microsoft.com/en-us/powershell/module/az.resources/remove-aztag`)

  - `New-AzTag` (`https://docs.microsoft.com/en-us/powershell/module/az.resources/new-aztag`)

- **Azure application programming interface (API)** (`https://docs.microsoft.com/en-us/rest/api/resources/tags`)

- **Automation deployments** via **Azure Resource Manager** (**ARM**) templates, or native and third-party tools (Terraform, Bicep, Chef)

- **Azure Policy**, via triggered automation

Let's now introduce the concept of a **business application**. This represents the application from the users' point of view, or simply *the service* that a user accesses or consumes, be it internal or external to the company.

For example, consider a simple **customer relationship management** (**CRM**) application (let's call it MyCRM) installed on Azure, with the following:

- One SQL Server database in PaaS

- Two backend application servers installed on two VMs

- Two frontend application servers installed on two VMs

- One Azure firewall

- One frontend application gateway

- One backend load balancer

In this example, MyCRM is the business application name.

Usually, a business application is related to one or many of these:

- **Application owner**: A person inside the organization that is in charge of the application service and life cycle

- **Organizational unit (OU)**: A unit or office inside the organization that is responsible for the application service and life cycle

- **One or more environments**: For example, production, quality, problem determination, development, integration testing, and so on

Each of the previous Azure resources has a role inside the application infrastructure. For example, we can identify three typical roles (or tiers, in standard terms), as follows:

- Database

- Backend

- Frontend

In our box-warehouse representation, you can ideally attach a label with some data to each component. For example, we can use at least three tags to help you identify the applications and the applications' components, as follows:

- Business application

- Environment

- Role

In our example, we would land on something like this:

Azure Cost Governance **RESOURCE TAGGING EXAMPLE**			
name	**BsnApp**	**Environment**	**Role**
Sql server – PaaS	CRM	Production	database
Backend servers	CRM	Production	backend
Frontend servers	CRM	Production	frontend
Application Gateway	CRM	Production	frontend
Load Balancer	CRM	Production	backend
Azure firewall	CRM	Production	frontend

Table 2.1 – An example of resource tagging: full list

A lot of companies have other environments—for example, a minimal quality environment to test deploys, perform code quality testing, and so on. In this case, following our example, we would land on this:

Azure Cost Governance			
RESOURCE TAGGING, MVP			
name	**BsnApp**	**Environment**	**Role**
Sql server – PaaS	CRM	Quality	Database
Backend server	CRM	Quality	backend
Frontend server	CRM	Quality	Frontend
Azure firewall	CRM	Quality	Frontend

Table 2.2 – An example of resource tagging: minimal list

You can add other tags related to areas such as these:

- The name of the software installed on the VMs (example: **Tomcat 8.1**)

- The granted service start-and-end time (example: *09:00-21:00*)

Let's continue our example by adding another web application (let's call it `MyWebApp02`) to our infrastructure, which shares some resources. The application relies on these components:

- Two frontend application servers, installed on two VMs

- One Azure firewall

- One frontend application gateway

Of course, for cost-saving, you think it's a good idea to share some resources—for example, the firewall and the application gateway.

Your tagging would then evolve into something like this:

Azure Cost Governance RESOURCE TAGGING, FINAL LAYOUT			
name	**BsnApp**	**Environment**	**Role**
MyCRM Sql server – PaaS	CRM	Production	Database
MyCRM Backend servers	CRM	Production	backend
MyCRM Frontend servers	CRM	Production	Frontend
Application gateway	CRM; WAPP2	Production	Frontend
MyCRM Load balancer	CRM	Production	Backend
Azure firewall	CRM; WAPP2	Production	Frontend
MyWebApp02 frontend servers	WAPP2	Production	Frontend

Table 2.3 – An example of resource tagging: final layout

As you can see, the shared resources have two business application **identifiers** (**IDs**) that allow you to know that the costs need to be shared between two business applications.

Tagging use maturity

Usually, as you start using the public cloud (either IaaS or PaaS) and have a bunch of resources, it's easy to maintain tags manually.

After some time, though, as your cloud infrastructure grows and the number of dedicated (used by one business application only) and shared (used by two or more business applications) resources grow as well, you may start experiencing delays or difficulties in applying the correct tags in a timely fashion.

Soon, you'll discover you need some automation to check the tags and report anomalies or automatically apply the tags.

The starting point of a good inventory and tagging strategy for large companies with on-premises and mixed cloud environments is, in my experience, a good **configuration management database** (**CMDB**) that should contain a complete inventory of all the *assets* (resources) in your infrastructure and that should allow you to define and *link* each resource to the correct business application(s). Having an accurate inventory, in addition, allows you to automate resource tagging.

With companies that have started directly in the cloud, this might not be necessary, as everything can be covered and automated with the Azure native tools (that is, management groups, blueprints, policies, automation).

You can automate resource tagging with the following:

- PowerShell (`https://docs.microsoft.com/en-us/azure/azure-resource-manager/management/tag-resources?tabs=json`)

- Linux AZ CLI (`https://docs.microsoft.com/en-us/cli/azure/tag?view=azure-cli-latest#az_tag_update`)

- API (`https://docs.microsoft.com/en-us/rest/api/resources/tags/create-or-update-at-scope`)

- Policy (`https://docs.microsoft.com/en-us/azure/azure-resource-manager/management/tag-policies`)

Management group policies for cost savings

Earlier in the chapter, we learned what **management groups** are. This is a key concept of **Azure cost governance**, as a correct configuration of management groups will allow for the following:

- Cost administrators for departments can have a partial view of the costs, depending on their privileges and according to the management group hierarchy.

- A few policies that can help the automation of cost governance can be enforced.

As per the official documentation, some built-in policies can help the cost governor enforce some rules, such as the **allowed VM size SKU**, the **add a tag to resources**, or the **allowed resource type** policies.

A very important policy is the **tagging** policy. Because we need tagging for cost control, the recommendation is to immediately enforce the tagging policy so that no new resource is created without proper tagging.

> **Important Note:**
>
> The **add a tag to resources** policy does not mean that the policy automatically assigns the right tag to new resources, since the automation cannot guess the final aim of that resource. You can still plan to automatically remediate missing tags via automation or you can add a *non-compliant* tag to help you add the correct tag later.
>
> If you are approaching **per tag** cost management and representation, this automation could allow you to implement an *unassigned* basket of resources that have been deployed but not yet *assigned* to a business application, and decide how to deal with it through policy automation, by automating or manually adding missing tags.

The remaining policies depend strictly on how you want to implement your cost governance process, and we will use them later in the book when using saving techniques such as **Reservations**.

For example, let's assume you have made a large reservation for a VM SKU that is therefore discounted at over 50% of its pay-as-you-go price. You want all users to be able to only use that SKU when creating a VM, therefore you enforce a policy of *allowed SKU* where only the reserved SKUs are listed.

> **Important Note:**
> When deploying VMs in an automated fashion, the deployment of any other SKUs will fail, and the log will show the specific error condition and policy that failed. If the deployment team has no access to policy management, you might need to establish a process where they send an alert with a log to the administrator to see if the policy should be modified. We recommend when enforcing such policies to inform your Azure users so that they will make sure to use the right SKU.

Earlier in this chapter, we spoke about horizontal scaling. Since this approach means that you will create new resources, you must ensure that they have the correct tags, and the last policy could help you automatically tag new resources inheriting tags from the parent resource group (if you organize your resource groups wisely, you will end up with a totally automated assignment).

Automation of cost control

In the previous section, we learned how to analyze costs manually and identify which screw you've to turn to optimize the cost footprint of an application, like a cloud watchmaker.

Just as every craftsman has a set of tools to tighten the right screw or loosen a bolt, you have a few powerful tools to operate on Azure resources, as follows:

- The PowerShell CLI
- The Linux AZ CLI
- The APIs
- Automation (ARM and third-party tools)

With these tools, you can integrate Azure resource configurations and actions directly in your batch jobs, and even schedule actions from your enterprise scheduler.

Question 12: *Do I need to have an expensive enterprise scheduler that natively supports Azure?*

No, it's not required. You can create your PowerShell or bash script and run it with the Windows Task Scheduler tool or with the Linux crontab, or work with Azure Automation and configure a workbook to run your scheduler.

Question 13: *Do I need to know a scripting or programming language?*

Yes! PowerShell and Linux AZ CLI let you create scripts. If you want to integrate through API calls, you need to know a little bit about programming languages.

The benefits of PaaS services: demand shaping

Cloud services (PaaS and SaaS) are very interesting from a cost governance perspective because they allow you to modify tiers and parameters in a very easy way, usually with one or two CLI commands.

Together with the three tools listed previously (PowerShell, Azure CLI, and API), these CLIs and APIs are very useful when you need to *shape* your resources during the day, week, or month.

Let's clarify this concept with an example: we will analyze an IoT Hub service receiving 3,000 messages per day, with a size of 200 bytes per message. With this rate, we could use even the Free Tier (assume 3,000).

> **Important Note:**
> Please note that the IoT Hub, as with many PaaS services and other resource types, can often only scale inside a tier—for example, you can scale from S1 to S2, S3, and vice versa, but not from S1 to B1 or to a Free Tier.

Let's assume that during the first week of the month, for some reason you need to receive 3,000,000 more messages per day, with a size of 0.4 **kilobytes** (**KB**). The message pattern is illustrated here:

Azure Cost Governance

IOT HUB MESSAGES PATTERN

period	messages per day	message size (byte)	min tier
1st week of the month	3,000 3,000,000	200 400	S1 or B2
other weeks	3,000	200	S1, B1 or even free tier

Table 2.4 – IoT Hub messages pattern

In a normal implementation, you'd have to size your IoT Hub to the maximum tier to manage the first-week-of-the-month peaks, but this would lead to a waste of money and resources in the non-peak hours.

Since the cloud is based substantially on a **pay-as-you-go** approach, it's very important to switch to a **pay-for-what-you-really-need** approach.

The usage trends can be analyzed through Azure Monitor logs or diagnostic logs, and then, according to the automation configuration, they can trigger an upgrade or downgrade runbook. In our example, you can create two scripts, as follows:

- `Iothub_upgrade_tier.sh`: This script will use an AZ CLI command to change the tier to `S2`, as illustrated in the following code snippet:

```bash
#!/bin/bash
my_subscription= "YYYYYYYY-XXXX-BBBB-AAAA-GGGGGGGGGGGG"
my_resource_group="ne-rg-iot"
my_iothub="inxneiot"
az iot hub update --subscription "$my_subscription" --
resource-group "$my_resource_group" --name
"$my_iothub" --sku "S2"
```

- `Iothub_downgrade_tier.sh`: This script will use an AZ CLI command to change the tier to standard `S1`, as illustrated in the following code snippet:

```bash
#!/bin/bash
my_subscription= "YYYYYYYY-XXXX-BBBB-AAAA-GGGGGGGGGGGG"
my_resource_group="ne-rg-iot"
my_iothub="inxneiot"
az iot hub update --subscription "$my_subscription" --
resource-group "$my_resource_group" --name
"$my_iothub" --sku "S1"
```

Go to the following link for an AZ CLI command-line reference: https://docs.microsoft.com/en-us/cli/azure/iot/hub?view=azure-cli-latest#az_iot_hub_update.

Schedule the scripts to run at the following times:

- 00:01 on the first day of the month: `Iothub_upgrade_tier.sh`

- 00:01 on the eighth day of the month: `Iothub_downgrade_tier.sh`

The total cost of the two approaches is summarized in the following table:

Azure Cost Governance	
IOT HUB TIERS COST	
approach description	total monthly cost
full month in S1 Tier	25$
1st week in S1 tier, other days in Free Tier	about 6$

Table 2.5 – IoT Hub tier approach and cost

Other examples of PaaS resources that allow demand shaping are provided here:

- **IoT Hub**: Changing tier for peak/non-peak hours.

- **Service hub**: Changing tier for peak/non-peak hours.

- **Azure NetApp Files**: Volume tiering (standard, premium, or ultra) and volume shaping (enlarging or shrinking your volumes based on performance needs).

- **Managed Disks**: Changing tier during non-peak hours, evaluating OS striping or mirroring, as in the previous example.

- **SQL Database**: Changing tier for peak/non-peak hours or switching to serverless (if you have a legacy application, you could still use this feature, automating the switch from general-purpose to serverless and powering off the application, in non-peak hours, then going back to general-purpose before starting the application).

- **Cosmos DB**: **Database transaction unit** (**DTU**) change for peak/non-peak hours.

- **Azure Backup:** Modeling retention times according to costs.

- **Managed Disk snapshots**: Implement an automation script that deletes, for example, every unattached snapshot older than 30 days.

- **App Service**: Change the App Service plan to a lower tier or share the same plan as much as possible in non-peak hours. This can also include Functions and WebJobs with a dedicated plan. You can also integrate those into your CI/CD to destroy the app services and the plans out of business hours and recreate them and redeploy the application from your repository to maximize savings.

> **Tip:**
> Serverless workloads using Azure services such as Functions, Logic Apps, and **Azure Container Instances** (**ACI**) in a pay-per-use approach represent the best and cheapest way to run highly scalable workloads without provisioning any specific plan or resource and are able to match exactly your workload requirement.

We've seen how tagging, policies, and automation can help us shape the costs of our cloud resources. In *Chapter 3, Monitoring Costs, Chapter 6, Planning for Cost Savings – Reservations*, and *Chapter 7, Application Performance and Cloud Cost*, we'll provide more examples of automation and—especially—demand shaping for PaaS services, as this is a trend that is growing at the pace of PaaS service adoption, and we predict some of these automation scripts might even be included in the Azure services at some point.

Summary

In this chapter, we learned how to define your cost constraints, split your logical cloud costs, and match the subscription hierarchy with how you want to charge those costs, by leveraging useful tools such as management groups, policies, tagging, and automation. The key takeaway is that you need to consider all cloud costs as fluid and try as much as possible to aim for scalable costs that can be matched, where possible, to the smallest billing increment possible for usage of the specific resources.

In the next chapter, we will learn how to monitor your costs progressively and continuously, which is the next logical step for defining the full cost governance process.

Questions

1. What are a Meter Category and a Meter Subcategory? And what is the finest-grained one that allows you to analyze specific dimensions?

2. What is a tag, why is it so important, and how can you apply it on Azure resources?

3. Are there resources in Azure that have only one meter subcategory? How do you deal with optimizing those services?

4. Why do you need to change your server vision from pet to cattle?

5. Which native tools can you use to automate actions and cost savings on Azure resources?

Further reading

- Management groups documentation: `https://docs.microsoft.com/en-us/azure/governance/management-groups/`

- Azure subscriptions limits and quotas: `https://docs.microsoft.com/en-us/azure/azure-resource-manager/management/azure-subscription-service-limits`

- Azure Spot VMs: `https://docs.microsoft.com/en-us/azure/virtual-machines/spot-vms`

- GitHub repository for built-in policies: `https://github.com/Azure/azure-policy`

- Ephemeral OS disks: `https://docs.microsoft.com/en-us/azure/virtual-machines/ephemeral-os-disks`

3
Monitoring Costs

In the previous chapters, we learned how to read, understand, and export cost details, as well as how to match your company's organization with how you want to see and charge Azure costs. Now, it's time to establish a solid monitoring process for your needs. This is the last step of preparing your cost governance process before you enforce saving policies and techniques, which we'll explore in the next chapter.

In this chapter, we will cover the following topics:

- Learning about and using Azure budgets and alerts
- Adopting Azure Cost Management Power BI
- Defining and configuring automation scripts for cost control
- Thinking about an architecture for a custom cost management tool

By the end of this chapter, you will be able to successfully monitor and control costs through dashboards and alerts regularly.

Technical requirements

For this chapter, you'll need the following:

- A computer with internet connectivity.

- Access to the Azure portal, along with a working subscription and the related credentials.

- Privileges to access Cost Management information (see `https://docs.microsoft.com/en-us/azure/cost-management-billing/costs/assign-access-acm-data`).

- To execute cost ingestion and splitting scripts, you'll need a host with at least **PHP 7.4** and a **MySQL** or **MariaDB** database.

You can find the code for this chapter here: `https://github.com/PacktPublishing/The-Road-to-Azure-Cost-Governance/tree/main/Chapter%203`.

Learning about and using Azure budgets and alerts

In this section, we'll learn how to view, export, configure, and automate how to handle billing data for budgeting and alerting purposes.

Azure Advisor

A good starting place for cost governance and optimization is the **Azure Advisor Costs** section. Azure Advisor is a sort of assistant that will analyze your current Azure subscription(s) and come up with smart suggestions on many topics, including cost, performance, reliability, and security. This information is automatically calculated and loaded for you in a friendly dashboard, and you can create alert rules based on Advisor's suggestions.

> **Important Note:**
> It takes a few days for Advisor to refresh cost information, so when you make changes, you might find that they are reflected in the tool with some lag.

Later in this book, we will learn how to make the most out of it. For now, the relevant part is that the information coming from Advisor can be automated by creating an **Alert Rule** that, for example, will send a message to the cost governor whenever there are high impact notifications on the cost category, as shown in the following screenshot:

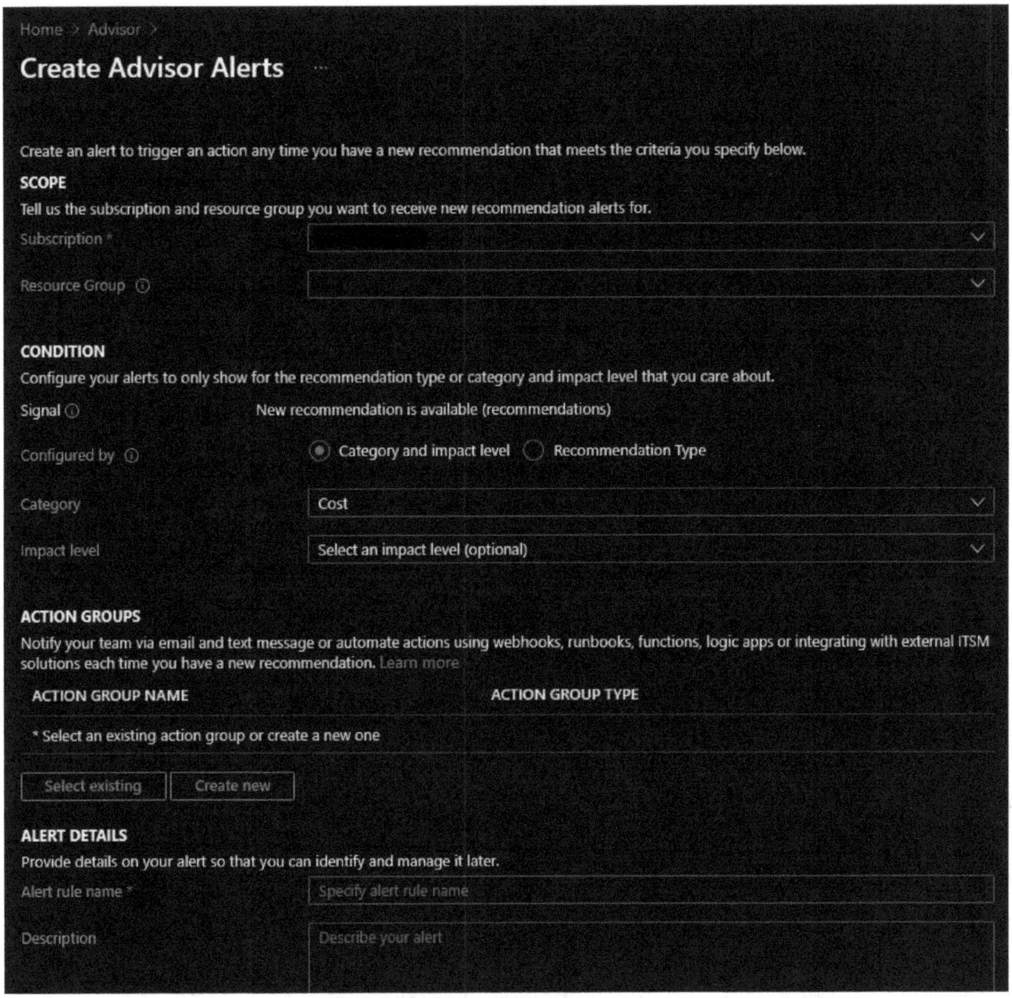

Figure 3.1 – Creating alerts for the cost category of Azure Advisor

Azure Advisor *should be your first stop* to get a view of your Azure spending, where you are wasting resources (and money), and all the possible recommendations for optimizing your resources' usage and spending. Hints are displayed for IaaS and PaaS resources, and features of this tool are added with *cloud* frequency and pace, so it will be easier to gather information about how to save money and optimize your spending directly from the Azure portal's **Advisor** page.

Dimensions analysis

Another important aspect in cost governance is understanding how to identify which component needs to be analyzed so that you can concentrate on the most expensive ones, or the ones you know you can work on with minimal effort (quick wins).

As we learned in the previous chapter, each cloud object generates costs based on different dimensions, including CPUs, GB of RAM, GB of storage, throughput for disks, the number of endpoint invocations for blob storage, messages per day, and message dimensions. These are only some examples of dimensions that influence the total costs of a cloud resource.

In Azure, these dimensions are as follows:

- **Meter Categories**: A high-level metric grouping; that is, a general way to unify similar *dimensions.*

- **Meter Subcategories**: This is the finest grain representation of *dimensions* (a Meter Category could represent one or more subcategories).

> **Important Note:**
> Meter Category and Meter Subcategory are suitable for a technical audience. If you need to export cost details for a non-technical audience (application maintainer, top managers, CFO, and so on), you need to translate it into a more business-related label. For example, IOPS for disks can be translated into *disk activity*, and message size for IoT hubs can be translated into *communication weight* or *payload size.*

Once you know that you need to analyze costs by Meter Category and Meter Subcategory (as introduced in the previous chapter in the *How cloud billing works* section), you only need to have a way to analyze your spending. We will learn about the complete process of cost governance in the following chapters.

When you know which deployed resources you need to concentrate on, why they are costing so much, and what dimensions you need to operate on to lower the cost, as stated at the beginning of this chapter, you need to know how the application works and how it consumes resources to optimize resources and reduce costs.

Let's clarify this concept with an example of the IoT Hub service. You can choose between two tiers (**Basic** and **Standard**). Each one has different dimensions, as shown in the following screenshot:

- Number of messages per day

- Message meter size:

Basic tier

Edition Type	Price per IoT Hub unit (per month)	Total number of messages/day per IoT Hub unit	Message meter size
B1	$10	400,000	4 KB
B2	$50	6,000,000	4 KB
B3	$500	300,000,000	4 KB

Standard tier

Edition Type	Price per IoT Hub unit (per month)	Total number of messages/day per IoT Hub unit	Message meter size
Free	Free	8,000	0.5 KB
S1	$25	400,000	4 KB
S2	$250	6,000,000	4 KB
S3	$2,500	300,000,000	4 KB

Figure 3.2 – IoT Hub tiers and pricing

There are different features too:

Feature	Basic	Standard / Free
Device-to-cloud telemetry	✔	✔
Per-device identity	✔	✔
Message Routing, Event Grid Integration	✔	✔
HTTP, AMQP, MQTT Protocols	✔	✔
DPS Support	✔	✔
Monitoring and diagnostics	✔	✔
Device Streams[PREVIEW]		✔
Cloud-to-device messaging		✔
Device Management, Device Twin, Module Twin		✔
IoT Edge		✔

Figure 3.3 – IoT Hub features by tier

If your application sends about 3,000 messages per day that are about 0.2 KB each, you can easily start with the **Free** tier (in the **Standard** tier), and you'll have a lot more features than the **Basic** tier, at no additional cost.

> **Important Note:**
>
> When dealing with Azure Services, as we learned in the previous chapters, the best practice is to use the online Azure Pricing calculator (which can be found at `https://azure.microsoft.com/en-us/pricing/calculator`) to inventory the prices for services and their sizes.

What happens if your application needs to send messages that are 0.6 KB in size? The **Free** tier is no longer enough for your application's needs, so you will need to switch to a **Basic B1** or **Standard S1** plan, just to add 0.1 KB of data.

At this point, the analysis will switch from the cloud/infrastructure point of view to the application development team: *is it possible to have the application send two messages under 0.5 KB instead of one message that's 0.6 KB?* This might be a little more complicated from a development point of view, and you may need to review the data model, but it will save you money in the short and long term.

Let's analyze another more common service: VM storage disks. In the following screenshot, we can see the pricing of Premium Managed disks, which will be needed for our exercise. You can find all the details of the Azure Managed disks service at `https://azure.microsoft.com/en-us/pricing/details/managed-disks/`.

	Disk Size	Price per month	1-Year Reserved Price Per Month	Max IOPS (Max IOPS w/ bursting)	Max throughput (Max throughput w/ bursting)	Price per mount per month (Shared Disk)
P1	4 GiB	$0.60	N/A	120 (3,500)	25 MB/second (170 MB/second)	$0.03
P2	8 GiB	$1.20	N/A	120 (3,500)	25 MB/second (170 MB/second)	$0.06
P3	16 GiB	$2.40	N/A	120 (3,500)	25 MB/second (170 MB/second)	$0.12
P4	32 GiB	$4.81	N/A	120 (3,500)	25 MB/second (170 MB/second)	$0.26
P6	64 GiB	$9.29	N/A	240 (3,500)	50 MB/second (170 MB/second)	$0.47
P10	128 GiB	$17.92	N/A	500 (3,500)	100 MB/second (170 MB/second)	$0.91

Figure 3.4 – Premium Managed disks pricing

If your application reads/writes to a 7 GB data file at 25 MB/second, you can easily choose **P2**.

What if your application needs to access the 7 GB data file at 35 MB/second? You may think the more logical tier is P6, but this is **7 times more expensive**. *But why not create a striped raid volume by using two or more smaller disks and striping them at the OS level?*

A RAID 0 striped volume with 2 x P2 disks will have 16 GB total with a max throughput of 50 MB/second, **at only 2 times the cost** (instead of 7).

Of course, this is a technical exercise that may be the responsibility of the technical team, but as the cost governor, you should at least be aware that this information will help you streamline costs in the end.

> **Important Note:**
> Azure Managed Disks is copied three times on the backend Microsoft storage, so you don't need to implement a RAID5 to protect your volume.

As you can see, if you identify your application's needs in detail, you can point out the best cost-effective implementation that will decrease the cloud spend and make it a standard for your company, for any other similar implementations.

> **Important Note:**
> When you implement a striped volume on your VMs, please consider all the implications: Azure Backup (which relies on disk snapshots) might not be used for application-consistent backups, because the snapshots of different disks are not atomical (there could be some seconds between one disk snapshot and the other, resulting in an unusable striped volume, with data loss). However, this is not an Azure issue: you need to verify all the implications and side effects before performing any configuration.

With this, you know the importance of analyzing every Meter Category and Subcategory of your resources. This will help you optimize the resource usage and costs on existing applications, as well as build new applications that are cost-optimized by default.

In the next chapter, we will work on a complete logic flow for this process, once we have identified and analyzed all the possible tools needed to address the cost governance process.

Useful representations

You've already learned how to use the Azure Cost Management tool to filter and group cloud costs. *But what are the most effective representations to be shared with the development teams or the application owners?*

Based on my experience, you may share the following:

- **Costs filtered per application, grouped by service name**: This helps you identify which component of the application needs a deep dive.

- **Costs filtered per application, service name, and grouped by meter subcategory (the most fine-grained dimension)**: This helps you identify which dimensions should be analyzed per service.

You can share two different types of representations, as follows:

- **Actual costs**: For example, the total cost sum of the last week or the last month.

- **Historical costs**: For example, a day-by-day costs representation of the last month.

A historical view is very important because it helps you identify a trend, a spike, or a sustained usage increment for a specific resource.

> Tip:
> When you represent the historical cost on a timeline, it's very useful to try to match the changes you see in the graph with the application changes, such as newly released features (*more resource usage or resources used in the wrong way?*), a widened scope usage (*increased number of users? X5? X10? X100?*), or increased data volume or retention.

Again, effective cost control is only feasible if you know the application and its evolution; otherwise, it will be extremely difficult to identify any incorrect infrastructure usage and avoid future extra costs promptly.

Scheduling exports

One of the first steps in monitoring Azure costs is creating a daily export of the cost information that you wish to receive at a glance. To do this, we need to open the **Azure portal** and navigate to **Cost Management**, then **Cost Analysis**, and then select the scope (subscription(s) or management group(s)) and define the view, as we learned in *Chapter 1, Understanding Cloud Bills*, in the *Export methods* section. This is the information that will be exported. By clicking on **Exports**, we can add a new scheduled export of the cost information.

We can identify the export with a friendly name, such as **Marketing**. The export name must be alphanumeric, without whitespace, and 3 to 64 characters in length.

The export type can be daily, weekly, monthly, or one-time. We recommend starting with a daily view so that you can start thinking of cost governance as a daily process and job.

For any scheduled export, we need some storage where the actual files will be saved. This can either be an existing storage account or container, or you can set up a new one dedicated to the exports.

Once you are happy with the changes, click on **Create** and the scheduled export will appear in your **Exports** pane:

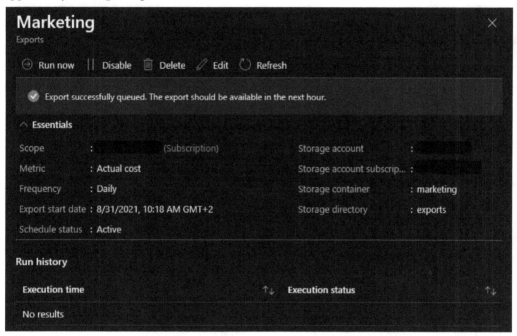

Figure 3.5 – Scheduling a billing data export

By clicking on the export, you can enable/disable, edit, and even run it manually. The export typically takes an hour or so before it's available in your storage account space.

Defining budgets and configuring alerts

Once you have learned how to read and analyze costs and how to properly identify owners and responsibilities for your organization, you can set up budgets and alerts, which will help you keep your running applications' costs under control.

In the Azure portal, navigate to **Cost Management + Billing | Cost Management** and select **Budgets**. At this point, you should select a **Scope** (that is, the subscription or resource group you wish to assign this budget to).

The page will display all the budget configurations you already have in place. To add a new budget, click on the + sign at the top left of the pane. This will open the **Create budget** pane:

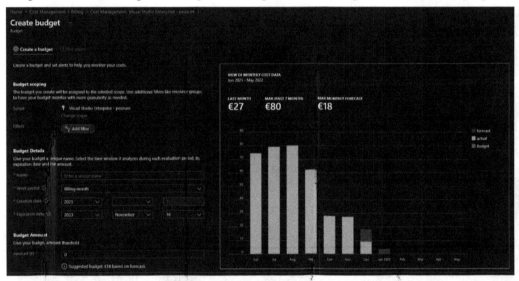

Figure 3.6 – Creating a budget

As you can see, the tool will automatically calculate the forecast for the selected scope and suggest a proper amount for the budget. You can enter a friendly name and determine its reset period and validity.

Once you are happy with your choices, click on **Next** to determine what actions this budget will trigger. First, we need to define the alert conditions; that is, what the trigger of the action is. This can be of the **Actual** or **Forecasted** type. We typically recommend an alert to be at 75% or 90% of the budget, but any amount can be entered according to your budget policy and based on the velocity of resource consumption. Once we have set the correct triggers, we need to identify the proper stakeholders and add them as recipients. Please be aware that the budget will only send a notification; the spending will continue.

> **Important Note:**
> Make sure you have whitelisted `azure-noreply@microsoft.com` in your email program so that alerts won't be marked as spam.

With that, we've learned how to check, display, and export billing information, along with how to add budget triggers that will send emails to the relevant stakeholders. In the next section, we'll learn what tools are available out of the box and what can be customized for your cost control monitoring.

Adopting Azure Cost Management Power BI

One of the monitoring tools that can help you deal with a cost governance process is the Azure Cost Management Power BI dashboard. Please refer to the following documentation to correctly install, configure, and deploy the dashboard. The Power BI app can be downloaded from `https://appsource.microsoft.com/en-us/product/power-bi/costmanagement.azurecostmanagementapp` and the relevant instructions are available at `https://docs.microsoft.com/en-us/azure/cost-management-billing/costs/analyze-cost-data-azure-cost-management-power-bi-template-app`.

This set of dashboards can only work for Enterprise Agreement customers and a Power BI Pro license is required for it to work. In the following sections, we will explore customized dashboards while using the **Free** tier.

There are a bunch of useful reports on the application, though we'll only highlight those we think are most relevant to the governance process. However, we encourage you to drill down into each of them and get the most out of all the neatly aggregated information provided:

- **Account overview**: This is a good starting point report, with a summary of information for charges and purchases.
- **Top 5 Usage drivers**: This report will display a summary of the top 5 spending Meter Category services.

The Power BI app for cost management provides a lot of useful dashboards and reports. We recommend that you specifically study and address the reports on **Azure Hybrid Usage Benefit** (**AHUB**) and Reservations since they will have a clear impact on the saving techniques that we will learn about in the next chapter.

Azure Hybrid Usage Benefit

One of the unique cost benefits of using Azure for Virtual Machines is Hybrid Usage, also referred to as AHUB or HUB. This is a licensing benefit that will allow customers who have already purchased Windows Server, SQL licenses, and RedHat and SUSE Linux subscriptions to save on the cost of those on-premises licenses when running on Azure. Additional information can be found at the following links:

- Windows Server: `https://docs.microsoft.com/en-us/azure/virtual-machines/windows/hybrid-use-benefit-licensing`

- SQL Database: `https://azure.microsoft.com/en-us/products/azure-sql/database/`

- Linux: `https://docs.microsoft.com/en-us/azure/virtual-machines/linux/azure-hybrid-benefit-linux`

> **Important Note:**
>
> The key takeaway of AHUB/AHB is that, in my experience, customers (especially large enterprises where this information might be scattered throughout departments) often *forget* to enable this benefit and waste money buying pay-as-you-go operating systems or SQL licenses that they already have. Enabling AHUB will only modify metadata, without any downtime, and can be fully automated.

It is important to be aware of the purchased and entitled licenses, and specifically, how you monitor their usage with the related Power BI report; that is, **Windows Server AHUB/ AHB Usage**. This report will display the AHUB/AHB situation for Windows VMs only. It will also calculate the benefits based on the VM's vCPU number, which will suggest better savings to be applied, for example, to bigger VMs.

> **Important Note:**
>
> A report showing a similar output for the other benefits is not out-of-the-box, but the information can be pulled and organized in a custom dashboard. To create a custom dashboard, you will need to use the **Azure Cost Management connector** (`https://docs.microsoft.com/en-us/power-bi/connect-data/desktop-connect-azure-cost-management`) in Power BI Desktop and add further information to create custom dashboards.

Reservations reports

Reservations, as we will see in detail in *Chapter 4*, *Planning for Cost Savings – Right Sizing*, and *Chapter 5*, *Planning for Cost Savings – Cleanup*, are a big part of the cost governance process, and there are a bunch of reports ready to use that help you handle and manage reservations in a timely fashion:

- **RI purchases**: This will display what you have purchased.

- **RI Chargeback**: This will display the reservations across each billing scope and region using amortized data.

- **RI Savings**: This will display, by reservation, the savings you obtained and even the pay-as-you-go cost of not using reservations.

- **VM RI Coverage (shared recommendation)**: This will calculate and display the recommendation for reservations by VM family, along with the recommended quantity and normalized size.

> **Important Note:**
> Depending on your pricing contract type (MCA, CSP, or EA), reservations might be listed differently and your overall ACR may not always include reservations.

We have seen how a pre-built Power BI dashboard can help you jumpstart your monitoring of cloud costs by providing ready-to-use and at-a-glance information on the top spending resources and the top impacting saving techniques. This tool has licensing requirements that must be considered before adopting it, though, and in the last section of this chapter, you have options for building a dashboard for free.

Now, let's dive into automation scripts that can adapt to your unique application map and help you govern your costs in a tailored and customized way.

Defining and configuring automation scripts for cost control

Azure automation scripts are a key element of every governance process, including cost governance. Automation runbooks can be scripted with the Azure graphical interface, PowerShell, and Python. For editing, we encourage you to read the documentation and choose something according to each advantage and limitation (https://docs.microsoft.com/en-us/azure/automation/automation-runbook-types).

> **Important Note:**
> Azure Automation can work on Azure, on-premises, and other cloud providers (for example, AWS) to automate tasks. Specific permissions are needed to securely access these cloud resources using the least privilege rule.

You can find detailed information in the official documentation at `https://docs.microsoft.com/en-us/azure/automation/automation-intro`, and tons of pre-created runbooks on the official GitHub repository at `https://github.com/azureautomation`, but the key elements you need to know about automation scripts, for our purposes, are as follows:

- **Scheduling exports**: Our billing information can be exported and sent via email, but with automation, for example, it can be transformed and sent to a custom tool that will then import it into its database.

- **Budgets and alerts**: An automation script can be triggered by an alert, such as by an automation account or by using **Logic Apps**, including the budget alerts, and can do as much as switch off resources when a specific budget has reached its limit (recommended for non-production environments only).

- **ARM template deployments**: An automation script can invoke Azure Resource Management templates with any type of configuration change you can think of, from scaling IaaS and PaaS resources and tiers to redeploying entire infrastructures based on your cost constraints.

- **Starting and stopping VMs**: This is explained at `https://docs.microsoft.com/en-us/azure/automation/automation-solution-vm-management`. There is also a new version that's currently in preview (more information can be found at `https://docs.microsoft.com/en-us/azure/azure-functions/start-stop-vms/overview`). We'll learn how to make the most out of this script in the upcoming sections.

- **Policy enforcement and subsequent actions**: For example, an Allowed SKU policy might trigger an action where you send the user an email stating that the only allowed SKUs are those that were reserved and hence are cheaper.

- **Invoking runbooks via webhooks**: This allows you to create an easy interface for all your on-demand and automated Azure Services scaling.

- **Enforcing Desired State Configuration (DSC) and Azure Policy Guest Configuration**: This can help you define a valid state configuration that is continuously monitored and acted upon, including what you feel is relevant for your cost governance. You can read more about these at `https://docs.microsoft.com/en-us/azure/automation/automation-dsc-overview` and `https://docs.microsoft.com/en-us/azure/governance/policy/concepts/guest-configuration`, respectively.

> **Important Note:**
> Azure DevTest Labs is a service that allows you to automate Azure non-production resources provisioning and makes it easier to control costs by providing pre-packages policies and automation for on/off VMs and services.

With that, we have learned that Azure Automation is a powerful tool that allows you to effectively orchestrate all your resources within Azure (and even on-premises and other cloud providers), although putting everything together, along with the dashboards and reports that trigger some of the cost decisions you have made, might not be an easy task.

In my experience, this proliferation of automation can lead to a specific need: that of a customized cost management tool that addresses exactly what your company needs to monitor costs and trigger reactions to specific events and situations. This is the topic of the next section.

Thinking about an architecture for a custom cost management tool

As we mentioned previously, when addressing automation scripts, all the automation and scripting will lead to you needing a unique tool where you can store and retrieve all your Azure cost information and make customized aggregations and elaborations that are different from the out-of-the-box dashboards provided by Azure.

Many third-party tools can fit this need, and we encourage you to discover them all and understand how they are programmed and if they match what your cost governance idea is.

Another way of dealing with this is to create a tool. This will give you a deeper knowledge about how Azure bills its resource usage, opening your mind to a new way of optimizing costs and helping figure out any hidden costs.

A cost management tool is composed of the following components:

- A database for storing and elaborating on cost information
- Automation scripts for exporting billing information from Azure and ingesting it into the database
- A user interface for representing cost information
- APIs or hooks for automation integration

Here's a schematic view of a basic architecture:

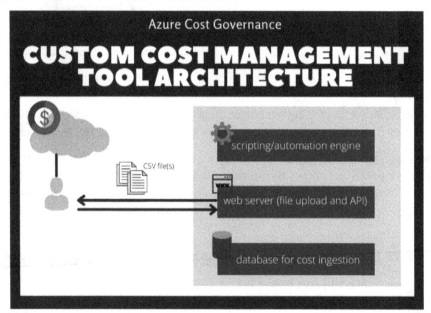

Figure 3.7 – Architecture for a custom management tool

Everything starts with downloading the monthly billing invoice (a .csv file) from the **Azure Cost Management** page:

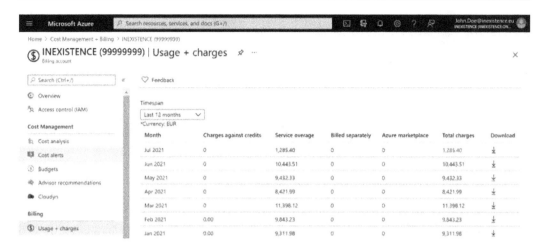

Figure 3.8 – Usage + charges report to be exported

Then, you must carefully study the fields and the information you obtain from that file.

At the time of writing, the `.csv` file contains the fields displayed in the following table. The highlighted fields are the ones you should focus on:

Fields in the .csv file			
BillingAccountId	BillingAccountName	BillingPeriodStartDate	BillingPeriodEndDate
BillingProfileId	BillingProfileName	AccountOwnerId	AccountName
SubscriptionId	SubscriptionName	Date	Product
PartNumber	MeterId	ServiceFamily	MeterCategory
MeterSubCategory	MeterRegion	MeterName	Quantity
EffectivePrice	Cost	UnitPrice	BillingCurrency

Fields in the.csv file			
ResourceLocation	AvailabilityZone	ConsumedService	ResourceId
ResourceName	ServiceInfo1	ServiceInfo2	AdditionalInfo
Tags	InvoiceSectionId	InvoiceSection	CostCenter
UnitOfMeasure	ResourceGroup	ReservationId	ReservationName
ProductOrderId	ProductOrderName	OfferId	IsAzureCreditEligible
Term	PublisherName	PlanName	ChargeType
Frequency	PublisherType	PayGPrice	PricingModel
CostAllocationRuleName			

Table 3.1 – All the fields in the CSV export file for billing

With this information in mind, you can start creating the first, simplified, version of your database:

- A *victim* table (let's call it az_billing_victim) that maps the .csv file 1:1.

- A *final* billing table (let's call it az_billing_details) that maps only the necessary fields for your tool.

- A tagging-resource table (let's call it az_res_businessapp) to map every tagged cloud resource to the correct business application and landscape (remember what we learned in the previous chapter in the *Tagging* section).

The content of the victim table should be deleted before every CSV load. This table lets you decouple the .csv file format from Azure (which we have *no control* over) and the final, historical billing table, which maps only the necessary fields we want to work on.

> **Tip:**
> When you transfer data from the victim table, az_billing_victims, to the final one, az_billing_details, you may start interpreting the Tags field to populate the tagging-resource table, az_res_businessapp. This will speed up the ingestion process, saving you time from having to reprocess az_billing_details a second time.

This is the *minimum* structure and workflow you need to start accumulating billing data and application/landscape relations:

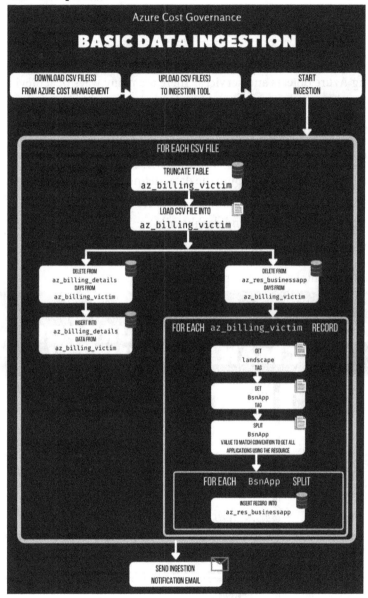

Figure 3.9 – Basic workflow of the data ingestion process

The first version of the data visualization can be just a web interface that exports simple HTML tables that you can analyze and maybe copy/paste into Excel files.

> **Tip:**
> You'll always need a simple, raw data visualization for debugging and cross-checking purposes, so you are probably going to maintain this textual representation for a long time.

Using the preceding depicted workflow, it is possible to build it on any sample ETL architecture using Azure objects and services. There are many ways of achieving the same goal, and many services can be used. One possible implementation is as follows:

- Data Factory to automate the ingestion process
- Storage account and blob storage to save/ingest the export files
- Azure Analysis services to combine the data, define metrics, and build the data model
- Power BI for presenting views and dashboards
- Azure SQL as the retention database

The following diagram implements the preceding points:

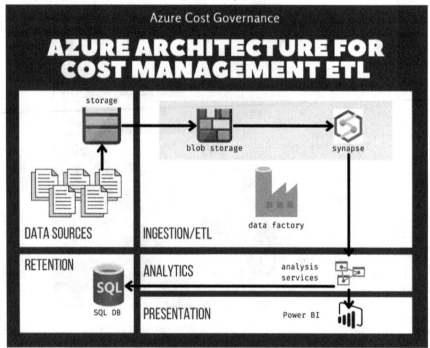

Figure 3.10 – Azure architecture for ETL

The next step is graphical data representation, where you need to identify the visual tool that fits your needs:

- Develop a web interface and use popular libraries to draw graphs on the web pages
- Use libraries to create static graphs in a batch automated manner
- Integrate the company's business intelligence tool

Before making any decision, you should consider the roles and skills of your team and of the people who build the GUI/graphs: if they have limited cloud or cloud-cost skills, and you're in charge of controlling and reporting cloud cost, it may be safer for you to implement an API layer to perform all the filtering, aggregations, loops, and so on before sending the data to Power BI.

For example, you can create a simple API to export JSON data for a custom Power BI dashboard: you can start with the free (but complete) Desktop edition, which lets you explore data, implement relations, create graphs, add filtering, and so on. On the other hand, if your team is well-versed in Azure objects, you can immediately work with the previous architecture and have an ETL process ready with a lot of additional perks, such as the analysis services, and maybe be add AI-powered queries to predict your spending.

You can also develop the workflow logic in your programming language. Let's look at an example that's been developed with PHP with a MySQL database.

The first script you need to run is from the `init_db.php` file, which can be found in the following GitHub repository: `https://github.com/PacktPublishing/The-Road-to-Azure-Cost-Governance/blob/main/Chapter%203/init_db.php`.

This script will create the target database and tables for the following purposes:

- Importing the billing `.csv` file from the Azure Cost Management tool
- Storing an official history of your Azure consumption
- Splitting the costs and storing the history of the split costs

Once you've created the empty structure, you could use the second script, from the `ingest.php` file, to load the `.csv` file from the Azure Cost Management tool. This script can be found here: `https://github.com/PacktPublishing/The-Road-to-Azure-Cost-Governance/blob/main/Chapter%203/ingest.php`.

It will take the input, as its first and only parameter, the path to the .csv file, which can be downloaded from the Azure Cost Management tool, and perform the following actions:

1. Load the .csv file into a temporary table to normalize some fields (such as the Date field).

2. Populate the historical table of the Azure costs from the imported .csv file to maintain the maximum possible granularity.

3. Populate the split costs table by inserting *N* rows for each row in the .csv file while using the *mathematical cost average* based on the number of business applications sharing the resource. For example, €12 worth of storage accounts shared by three apps will generate three rows in the split costs table with €4 each.

> **Important Note:**
> To allow the script to *know* which business applications share the resource and the related environment/landscape (for example, Production, Test, Quality, and Pre-Production), you need to define two tags on the Azure resources.

For example, for resources used by one business application only, use the following code:

```
Tag name  :  BsnApp
Tag value :  |Application1234|
Tag name  :  Landscape
Tag value :  Production
```

If the resource is shared between multiple business applications, use the following code:

```
Tag name  :  BsnApp
Tag value :

   |Application1234|Application9876|Application6547|
Tag name  :  Landscape
Tag value :  Quality
```

If you ingest the .csv file of the previous month on the fourth day of every month, you'll end up with the following being updated:

- Historical data for dashboarding and monitoring
- Historical split costs trend at the business application level

You can also start building up a dashboard to monitor and share details about the following:

- How much each application spends across time
- The type of resources and SKUs used by each application

You can also use the side effect of having a strong tagging policy to extract useful information for a CMDB, business impacts, or documentation purposes.

Translating for the non-technical audience

As you may remember, I previously hinted that it may be tough to explain all the technical aspects behind cloud management , as well as all the demanding application resources, to non-technical people. Concepts such as Meter Category (**Event Hub**, **Sentinel**, and **Azure Cosmos DB**) and Subcategory (**Azure Purview** scanning/data map or a Basic/Standard/Premium/Isolated plan) can be clear to those who designed the application(s) and manage the cloud infrastructure but are probably new to an executive or a cost controller.

For a non-technical audience, the first stop would be an Excel file with the exported data. This would be used as a Power BI tool for business executives to execute pivot tables, remap fields, and even hide fields that are not relevant to an audience. You would still need to work with your financial controller to map the KPI's expectations to the Meter Categories and Subcategories coming from the Azure Cost export.

> **Tip:**
> One of the fastest ways of getting non-technical people on board is involving them in the Azure fundamentals, or maybe even asking for formal training and certification (AZ-900). This will help them quickly absorb the technical jargon and help the translation work go smoothly.

So, you need to calibrate the cost representation according to your audience:

- Representing the actual costs (for example, the current month).
- Representing historical costs (for example, the last 12 months or all the months).
- Displaying any comparison between the months or years to show how the consumption changed across an appropriate period.

- Relabeling the technical metrics and dimensions so that they have more understandable labels, as follows:

  - *general blob block* to *object storage*

  - *ms series* to *memory optimized vm*

- Trying to tell the whole story: tech people may remember what happened to a resource or why you resized a disk, but this may be unknown to your management and should be explained (upon request).

- Representing costs split by business concepts (application name, landscape, and so on).

> **Tip:**
> Don't worry about merging costs due to Category and Subcategory relabeling. Explaining the difference between a managed and an unmanaged disk may be confusing and be a waste of time during a business meeting; it may provide unnecessary information when you're sharing costs with a non-technical audience. You need to be ready to share drill down(s) and technical concepts but overall, please try not to over-complicate your explanations.

You may implement the translation by simply adding a couple of tables (that is, `az_metercat_relabels` and `az_metersubcat_relabels`), in which you note one or more additional labels: you may need a couple of different translations for the cost controller, managers, and C-Levels audience. The next step is joining these tables with the `az_billing_details` columns.

> **Important Note:**
> Please do not forget to relabel any *new* Meter Category or Subcategory; otherwise, you will end up merging different cost types or having *unlabeled* costs!

Now, let's expand our workflow by adding some steps to check if, in the Azure Billing file, there are new Meter Categories and/or Meter Subcategories that need relabeling (you should set it to inform you by mail of any new cost labels that need to be addressed):

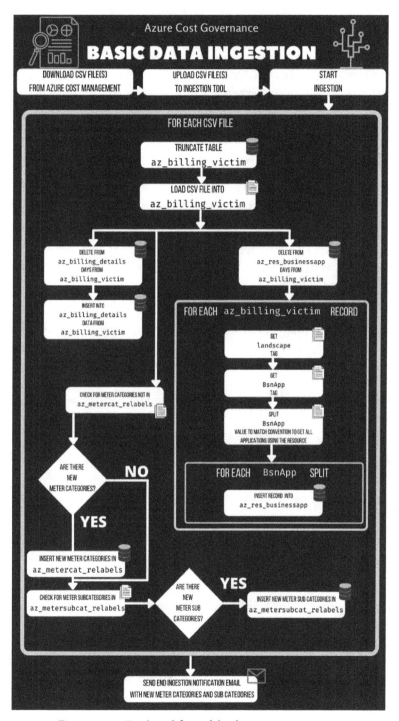

Figure 3.11 – Final workflow of the data ingestion process

With this example, we were able to define a solid workflow for ingesting all the billing information. You might find situations where this can get more complex than our example, and we encourage you to follow our recommendations for the script to be reliable and predictable, as well as flexible enough to be able to change it, should the cloud provider decide on minor and major schema substitutions.

Mind the gap between cloud costs and your cost control platform

If you're wondering if you can have *everything* under control, the answer, unfortunately, is that you cannot. Cloud costs formats, labels, and measure units may change very quickly, and the cloud provider typically does not inform you of this before applying changes.

In my personal experience, I have seen changes in the fields position, the decimal separator, the date format (from ISO to US and back), services measure units, and mixed string fields enclosed by quotes, and in some cases, even bigger changes such as aggregating Azure Plan IDs for partners without specific user notice.

So, you need to understand that your custom ingestion workflow relies on something that you don't fully control – that is, the raw data from the cloud provider – and that you need to carefully develop the integration and import steps to ensure that your database is the single source of trust for cost queries. Imagine importing a `.csv` file with the date in the wrong format, without parsing and validation: you'll end up with an unpredictable mess in the daily costs. *The key takeaway is that you need to account for these changes in your workflow and be prepared to quickly adapt your workflow and tools.*

Since you have access to both your custom cloud cost control and the official Azure Cost Management tools, a good practice is to check the same value on both sources, to make sure everything has been imported correctly (for example, the total monthly cost should be equal on both tools).

Summary

In this chapter, we learned how to automate most of our billing and cost control information by using scheduled exports, Azure Advisor, Azure Budgets, and Azure Alerts. We explored using the pre-packaged Power BI app for cost management and learned how powerful Azure Automation can be for cost governance. Finally, we learned how to think of a customized tool that can be tailored to our company's needs. Laying out a custom cost management tool is the foundation of a cost-saving strategy, as we'll explore in the next chapter.

You now know how to read, interpret, extract, and monitor your cloud costs, and can create a governance tool. The next few chapters will dive deeper into every cost-saving option so that your full cost governance process can make the most out of all the possible saving techniques.

Questions

1. Can you schedule when Azure cost information should be exported?

2. What is Azure Advisor?

3. What is AHUB? Why is it important?

4. How can you monitor your Reservations capacity and usage?

Further reading

- Azure Advisor documentation: `https://docs.microsoft.com/en-us/azure/advisor/`

- Azure Hybrid Benefit FAQ: `https://azure.microsoft.com/en-us/pricing/hybrid-benefit/faq/`

- Azure Reservations documentation: `https://docs.microsoft.com/en-us/azure/cost-management-billing/reservations/save-compute-costs-reservations`

- Azure Automation documentation: `https://docs.microsoft.com/en-us/azure/automation/`

- Azure Devtest Labs: `https://docs.microsoft.com/en-us/azure/devtest-labs/devtest-lab-overview`

Section 2: Cloud Cost Savings

In this section, you will understand and use cost-saving options such as reservations, and techniques to downsize, clean up, and switch off/on your Azure resources, along with offering on-demand solutions where possible for them. We have a dedicated chapter for right-sizing, another for clean-up, and one for reservations.

We'll understand how to keep the right balance where possible between performance and spending, and will learn how to identify, automate, and delete unused resources. This section will also explain how to properly set up a governance process to handle reservations, from understanding the model options to successfully handling a full capacity strategy of reserved services.

This section comprises the following chapters:

- *Chapter 4, Planning for Cost Savings – Right-Sizing*
- *Chapter 5, Planning for Cost Savings – Cleanup*
- *Chapter 6, Planning for Cost Savings – Reservations*

4
Planning for Cost Savings – Right-Sizing

In the previous chapters, we have provided all the relevant elements to be able to understand, export, analyze, and monitor your Azure cloud costs. It is now time to learn how to save money by identifying all the possible actions that can reduce your spending, while still leaving a good balance between the performance and reliability of your applications.

This is the first chapter dedicated to cost saving, and it starts, as mentioned often in the first three chapters, with changing the long-used **information technology** (**IT**) culture of overallocating servers to a strict, Lean, and Agile mentality of using *just the right amount of resources*.

We will learn how to tackle right-sizing for **virtual machine** (**VM**), storage, and **platform-as-a-service** (**PaaS**) services, and then prepare internal policies for on/off and on-demand automation. Lastly, we'll finish this chapter with a non-exhaustive list of logic and workflows to help you keep your resources and your costs at the right size and usage. **Right-sizing** is all about finding the perfect balance between performance and cost and periodically enforcing your cost/performance policy to get the most out of your cloud spending.

You'll also learn about sample logic for cost control with right-sizing, where we'll guide you to analyze, control, configure, and automate some of the cost-saving techniques we learned.

These will be the main topics of this chapter:

- Understanding right-sizing
- Enforcing on/off policies—scheduled and on-demand
- Sample logic for cost control

Upon completion of this chapter, you will have a clear understanding of the most used best practices and techniques for controlling cloud costs, along with practical examples of how to apply the newly acquired knowledge.

Technical requirements

For this chapter, you'll need the following:

- A computer with internet connectivity
- Access to the Azure portal
- Privileges to access **Cost Management** information (see `https://docs.microsoft.com/en-us/azure/cost-management-billing/costs/assign-access-acm-data`)

Understanding right-sizing

As has been mentioned, and emphasized many times since *Chapter 1*, *Understanding Cloud Bills*, cloud computing is not about overprovisioning, and right-sizing is probably the most important aspect of controlling cloud resources and costs. The impact is not simply saving money: in many cases, there is a balance between performance and spending and, more specifically, between meeting your *internal customer* **service-level agreements** (**SLAs**) efficiently. You need to find this balance to keep both your application managers and cost controllers happy. Let's dig into this further by dividing the scope of application—where the first step to right-sizing starts, of course—from your **infrastructure-as-a-service** (**IaaS**) resources.

Right-sizing is the process of continuously finding the right balance between resources' performance (**RAM**, storage, **CPU cores**) and the lowest possible cost for those resources. This means performing an assessment of your resources on a periodic basis, which will lead to three main actions, as follows:

- **Upsize scaling** of your resources vertically (VM sizing, storage, tiers, and so on)
- **Autoscaling** horizontally and adding nodes

- **Downsize scaling** (VM sizing, storage, tiers, and so on)
- **Switch-off**—that is, trying to limit the usage of cloud resources to just the time these are really used

> **Important Note:**
> In some cases, we will learn that right-sizing might not be the correct fit for a workload. There can be many reasons— software requirements in terms of memory or processor, or in some cases, no maintenance window can be used to downsize resources, or resources and services cannot be turned off or scaled down (such as networking services), and so on. As a result, you may find yourself in a frustrating position of using overallocated resources by design: don't worry, though, as there are ways of saving money even in this uncomfortable position, as we will learn in the upcoming chapters.

Choosing the right balance between performance and cost is also an exercise that depends on the type of cloud workload we are working on: IaaS, storage, and PaaS, which we'll cover in the next sections.

Right-sizing for IaaS

If you have migrated from an on-premises situation, chances are (from my experience) that having worked with a few large migrations of global enterprises (with numbers varying from 1,000 to 10,000 VMs), you are using 30-40% of the compute power that you bought. This typically happens because sizing servers in a private data center means buying hardware that you will probably keep for 3 to 5 years (sometimes even more); hence, overprovisioning has been a common practice and even a requirement due to long procurement times, for many years.

Additionally, systems are sized for *peak performance requirements*, where this peak may only be hit one-two times a week. Problem is, when you migrate your workloads and make a one-to-one match with your cloud resources, you end up with a virtual data center that is overprovisioned and overspending for no good reason.

A good starting point to understand if you have overprovisioned resources is the **Azure Advisor** page in the Azure portal. From the main **Advisor** page, you can see **Cost** recommendations such as *right-sizing* or *buying reservations*, as illustrated in the following screenshot:

Figure 4.1 – Azure Advisor right-sizing suggestions

If you click on **Right-size or shutdown underutilized virtual machines**, you will get a per VM details list, which you can also export in .csv or .pdf format, which is useful for building up your business case with right-sizing and presenting all the findings. You can see this list in the following screenshot:

Figure 4.2 – Azure Advisor right-sizing detailed view

As you can see from the preceding screenshot, **Advisor** will list all the VMs that have a CPU usage below a specific percentage (the default starts at 5%, but this value can and must be changed with time). This report tells you that several resources in your virtual data center have average use of a CPU below a certain threshold, which means they are practically idle.

This reasoning pattern prompts some decisions that will vary according to your organization's policy and technical requirements. Here are some things you may want to think about:

- Pick a VM or group of VMs in the same resource group, or belonging to the same application:

  - If they are non-production, you might decide to downsize in the first available downtime slot.

  - If they are production, you need to contact the application manager and understand if there are specific requirements and constraints that prevent the downsize. For example, a larger VM might be needed for the number of disks or memory that the application requires. You need to make these checks before proceeding to any downsizing since they can have an impact on your SLAs and performances.

- Once you are happy with the lower size of one or more VMs, then you can schedule the moment of downsizing. If you plan to do this at the end of the month, this will give your cost controller an easier view of monthly quotas.

The VM and storage costs are typically the main chunks of spending of every customer, as you might check and evaluate with the *Azure Cost Management – Cost Analysis* reports that we learned about in *Chapter 3, Monitoring Costs*. Let's progress with our process by addressing the second largest spending item in the list: storage.

Disk right-sizing

Storage is typically also a big part of cloud spending, especially for customers who have a large IaaS footprint. There are several types of managed disks that reflect different performances, features, and, most of all, pricing, as follows:

- **Ultra-disk**: These are the fastest and most performing disks and, therefore, unfortunately also the most expensive. In addition, the pricing directly depends on the **input/output operations per second** (**IOPS**) and throughput that your application will require. My recommendation is to define exactly what performance (these disks were designed to address workloads such as *SAP HANA* or *Oracle* databases) you need from the storage before committing to Ultra-disk.

- **Premium solid-state drive (SSD)**: These are the most used since they provide good and solid performance and pricing. For cost purposes, however, most of the workloads using premium disks can easily be downgraded to standard SSDs or **hard disk drives** (**HDDs**).

- **Standard SSD**: Great price-performance ratio, although a bit limited on the IOPS and throughput. Microsoft recommends using them for **development** (**dev**)/test and light workloads.

- **Standard HDD**: This is the most basic tier of disk storage. It can be used for backup and non-critical workloads with sporadic access.

In addition, you can now share your managed disks with an *Azure disk pool (preview)*, allowing your application to access your managed disks from one single **iSCSI (Internet Small Computer System Interface)** endpoint. This feature is in preview (`https://docs.microsoft.com/en-us/azure/virtual-machines/disks-pools`) and will only work with pools of Ultra-disks or Premium/Standard SSDs—you cannot mix the two types.

> Important Note:
> Azure disks can be reserved as well, bringing down the cost by reserving increments of 1 disk unit for a 1-year span. We will learn about reservations in *Chapter 6, Planning for Cost Savings –Reservations*.

There are several options to keep managed disks at the right size, as follows:

- If you overprovisioned a Premium disk only for IOPS/throughput, you can change the disk size without stopping the VM, as much as twice per day (refer to the following for more details: `https://docs.microsoft.com/en-us/azure/virtual-machines/disks-performance-tiers-portal`), so during the day, when load is high, your disk can be a bigger size with related performance, while at night you can downsize to a smaller footprint.

- If your workload has spikes that affect the disk performance, you can enable disk bursting and only use additional resources when needed. See `https://docs.microsoft.com/en-us/azure/virtual-machines/disks-enable-bursting` for more information on this.

- If your workload has an on/off pattern, you can change the storage type to standard HDD when the VM is off, and change it back to Premium when the VM is on again. See `https://docs.microsoft.com/en-us/azure/virtual-machines/linux/convert-disk-storage` for more information on this.

We encourage you to read the full Microsoft documentation on what type of disk to use for your workloads, available here: `https://docs.microsoft.com/en-us/azure/virtual-machines/disks-types`.

> **Important Note:**
> Before resizing or changing the tier of a disk, you should make sure about its impact on the application, whether the VM supports both disk types, and any related extra cost triggered due to resizing or changing. For example, resizing to a disk that has transaction charges might incur extra fees when the disk is used, even if for a small amount of time.

The key takeaway is that you should aim to size your storage as much as possible according to your current needs, resisting the on-premises mentality of keeping extra space for growth, as the extra space in the cloud can and should be easily allocated only when you really need it. Of course, keeping a buffer of additional storage for organic growth is still a best practice but it should be measured according to real growth, and not some future extra plans where you might end up paying 20% of additional storage for no good reason.

> **Important Note:**
> Storage size is charged regardless of consumption, so if you decide to allocate a P30 disk (Premium 1,024 **gibibytes** (**GiB**) disk) while using half the space, you are still charged the full 1,024 GiB of storage!

Other reasons that may lead to storage overprovisioning can be any of these:

- Unassociated disks (for example, from VMs that were deleted)
- Aged snapshots
- Migrations of the wrong disk size—for example, a migration tool that converts a disk to 1,025 **gigabytes** (**GB**) will have the result of a 2 **terabytes** (**TB**) charge!
- Windows images and Hyper-V images can consume a lot of disk space, especially if the Hyper-V images were converted from VMware vSphere.

We will dig deeper into wasted resources in *Chapter 5, Planning for Cost Savings – Cleanup*, which is completely dedicated to cleaning up unutilized resources that only waste money. Let's now move on to the right-sizing of PaaS resources.

Right-sizing PaaS services

Many customers are starting to see the public cloud as a development platform, migrating their application to modern, cloud-native architecture and using cloud PaaS services that cut lots of effort from the development and operation times. However, just as for VMs, you need to properly choose the correct tier of your PaaS service to avoid extra costs and/ or performance that are not aligned with your business needs.

> **Important Note:**
>
> To be able to reconduct every resource to the right application and application manager (or owner), you should plan to have a **configuration management database (CMDB)** or a setup to use the Azure native assessment and discovery tools, which allow you to track down—at the minimum—the application, its owner, and the used cloud objects.

An additional complication of using PaaS services is that they often have different tiering that depends on how the service was built or on how the application uses it. This makes it hard for an operation team to assess and downsize a PaaS service since you will need to know exactly what types of requirements your application has.

The starting point for defining the right tier is, of course, to assess the memory and CPU usage of the application and determine how changes will impact the overall performance. A subtle problem comes when you are using the cloud platform service of your choice but do not realize you are pushing it at 100% or more.

Most services have a *throttling* mechanism that will refuse more workloads when a tier reaches its limits. However, many other services typically have a soft limit, and therefore our strong recommendation, especially for business-critical applications, is to enforce a good monitoring tool that will allow you to constantly assess and optimize the performance tier of the platform. Many PaaS services support autoscaling, but this is not an immediate operation and there are limits to how far you can scale at once: you will need to plan for your workload type in production and when headroom is necessary to support expected workloads.

> **Important Note:**
>
> *What happens when you downsize a PaaS tier?* The problem with downsizing a PaaS tier is the fact that you are not simply scaling down resources but, in many cases, are switching to a lower tier of product, and you might lose critical features that will require you to keep a higher tier even if the performance is great.

In this section, we have learned about how to right-size VMs, their attached storage, and the tiers of PaaS services. As mentioned earlier, this might not always be the correct—or, sometimes, feasible—approach. When downsizing is too complex, we will try other means of cost savings, such as—as we will learn in the next section—scheduling your workloads to run only when used.

Enforcing on/off policies – scheduled and on-demand

Since you're reading a cloud cost-optimization book, I think you may be one of the *turn off the tap when you don't need water* club members.

The same best practice could and should be used with cloud resources: if you don't use a service, database, or VM, then that resource should not be running without purpose.

Usually, we can identify two types of on-off policy, as follows:

- **Scheduled**: We use this policy when you set a *power-on* time and a *power-off* plan, and it's always the same.
- **On-demand**: We use this policy when the resources are always *powered off* and the user starts them up only when needed.

Of course, the on-demand approach is the best, when speaking in terms of cost control, because you use the cloud resources only when needed, and for the right time. The scheduled approach, instead, may leave the resources on even if unused.

Non-production environments (test, development, quality, pre-production, quality, and so on) are best suited to an on-demand approach. Production environments are typically always-on, to ensure a proper service level, although in some cases, they might take advantage of a scheduled approach.

At first, you may think that the only benefit of shutting down services when they are not used is limited to the billing costs generated by those resources. But if you bear with me for a while, I will demonstrate that this is not entirely true and that the whole infrastructure could get some additional benefits, such as optimizing reservations, saving on resources that are linked to the primary target of an on-demand policy, and even saving license costs, as we will learn in the upcoming sections.

Saving on reservations

Reservations, as we will learn in *Chapter 5*, *Planning for Cost Savings – Cleanup*, Azure backend systems may reassign hourly the discount to any *compatible* resources (for example, a different VM of the same family), as per the flexibility option.

If you flagged flexibility in your purchase when you shut down (that is, VM deallocation, as per the VM states we learned about in the previous chapters) one or more resources you originally intended for reservation, the discount will automatically apply to another running resource of the same group, as follows:

- **Powered-off** (reserved) resources generate zero (or near zero) costs.
- **Running** (unreserved) resources will benefit from the reservation that is freed up on the powered-off resources.

Induced right-sizing

When you stop a resource, all the attached (or related) resources can be right-sized since they're not used anymore. One simple example is VMs and disks: we learned that, in a cloud environment, the disk storage qualities (HDD, SSD, Ultra disks) have different price ranges. If your application normally needs Premium or Ultra disks when running, once powered off, these will no longer be used.

You may still decide that you want to keep the disks (hence, the storage cost) because it's an on-off exercise or simply a grace period for deleted VMs. As a practical example, you can imagine downsizing all the disks that are attached to powered-off VMs to Standard HDD and paying the minimum possible price. Before starting up the VMs, you will need to revert the disk tier back to the original one (to ensure correct performance when the VMs are running).

Here is a sample workflow to start/stop VMs from extending the saving with disk **SKU** change (an **SKU** is the price and feature tier of every cloud resource):

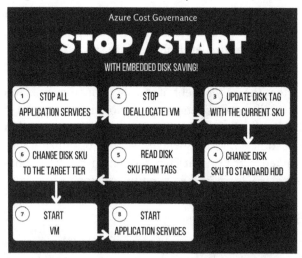

Figure 4.3 – Application start/stop script with disk saving

The example on the disk tiers teaches us that there may always be extra costs, which we can call to be inducted, that are a result of an apparently seamless change that can invalidate our savings efforts. In the same way, we may have inducted savings as a result of other operations.

A couple of other examples of inducted savings may include these:

- **Log collection**: These are active resources that send logs to a collector that generates costs on network, storage, and log ingestion. If the resources are stopped and deallocated, you won't have such costs.

- **Planned maintenance**: Having a time window in which some resources are down (and declared down to the user) allows you to plan maintenance windows (Azure maintenance or standard operations maintenance) without having to deal with the users for scheduling.

- This is not an Azure or Cloud Direct saving, but at the end of the year, cloud operations will save **full-time equivalents** (**FTEs**)—namely, the people working on those workloads—as a cloud deployment typically incurs lower operational costs due to the fact you are not managing the hardware and data center part, and in complex infrastructure, this may save *a lot* of FTEs.

These are just some examples to better understand what inducted costs are and the importance of evaluating everything in your applications, and not only the resources involved in *serving* the application.

Licenses

If you buy resources in a **pay-as-you-go** licensing model, having them running only for a fraction of the day will save hours and hours of license costs.

This is very important for the total project spending since usually, product licenses are bought *upfront*, but if you carefully plan your on-off strategy, you could switch non-production environments to a **pay-as-you-go** approach with tight schedules, to be sure you end up spending less than buying them upfront.

It's still worth mentioning that for Windows and **SQL Server** licenses (and every other license type covered by this benefit), if you have enabled **Azure Hybrid Use Benefit** (**AHUB**) (as mentioned in the earlier sections), it will cut the license cost of the workload.

Backup

Some resources cannot be backed up using Azure Backup (for example, application data files or databases on which Azure can't guarantee consistency) and require specific software to back up those items on different storage. Usually, specific third-party backup software is licensed by *backed-up data size*. So, the more data you back up, the more expensive the license is.

But powered-off systems are, by definition, consistent, since there's no data access or concurrent data modifications. Therefore, you can switch to Azure Backup, for example, freeing up traffic and licenses from your specialized software.

> **Important Note:**
>
> Please make sure that while backing up your system, once it's powered off through snapshot or copies, ensure the overall application's data consistency and the complete restore for each workload. In any case, I suggest verifying this with the software supplier.

Network storage

If your application uses **network-attached storage** (**NAS**) that can change tiering (and, therefore, cost), please consider limiting the performance and space of it whenever your application is powered off.

Azure NetApp Files (**ANF**) is an example of NAS that can be hot-resized (the tier can be changed with no downtime) and automated via **API** or AZ **CLI** commands to shrink the capacity pool and volumes when not in use.

The previous VM stop/start workflows may evolve as in the next diagram, where, in addition to resizing your managed disks according to your usage, you also increase and decrease ANF volumes (always tagging them properly with the latest change) to get a proper demand shaping. This behavior is not implemented yet in ANF but might be in the future, reducing all our diagrams to one simple flag of *demand shaping*.

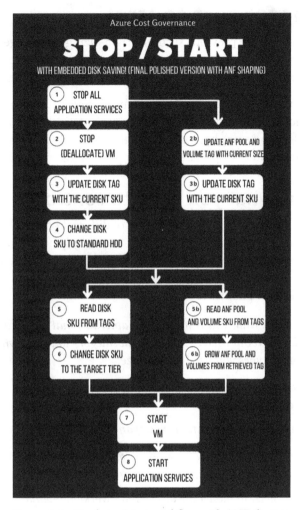

Figure 4.4 – Final stop/start workflow with ANF shaping

In the preceding diagram, we have tried to define the outcome of our start/stop workflow. You are obviously welcome to make it more complex according to your company's requirements; this should be a **minimum viable product** (**MVP**) to allow you a safe stop/start for your cloud applications.

In this section, we have learned what are the easiest and quickest targets for cost savings related to right-sizing. We have also learned how to prepare and enforce off/on and on-demand policies. We will put everything together with some real-life examples of cost control logic in the next section.

Sample logic for cost control

As you may have guessed in the previous sections, right-sizing can be overwhelming when dealing with large complex organizations and/or projects. In addition, there isn't a one-for-all approach that, once applied, saves you tons of money with little effort.

Of course, policies help to maintain control, but you need to adapt Azure resource SKUs to your workload demand, and this cannot be achieved with a static, one-time-only action. In this section, you'll understand some basic principles that may help automate tasks to control costs using AZ CLI commands, but you are free to implement them as you prefer.

Understanding what you need to control

Usually, in your cloud infrastructure, you have a lot of different objects, but don't necessarily have to create automation: some objects may simply not need intervention. So, let's start posing some questions about your workload demands that, indeed, need attention, with sample answers. The answer to the following questions is specific to your organization and application policy:

- **Question 1**: *Based on what you learned in Chapter 2, What Does Your Cloud Spending Look Like?, about Meter Category/Subcategory analysis, what are the resources or resource types you need to optimize?*

  For example, Azure managed Premium disks, networks, VMs of a specific family (for example, E or M family).

- **Question 2**: *What are the applications you want and can tackle?*

  Based on the top *N* most expensive resources, you can identify the most spending architecture/application (for example, SQL Server in PaaS of the MyCRM application).

- **Question 3**: *If you need to model the application workload demands over time, are you able to represent them in a graph or in a timeline?*

  We can use some examples of application usage patterns that may help you to understand cost control. They are provided here:

  - **Working days peak**: Your application uses a lot of resources systematically from Monday to Friday. Here is a screenshot to demonstrate this:

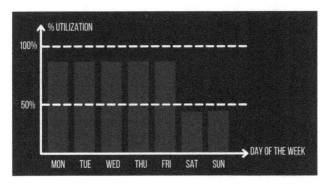

Figure 4.5 – Application usage with a peak on working days

- **First days of the month peak**: Your application consumes resources only in the first few days of the month. After that, it is completely idle or unused, as illustrated in the following screenshot:

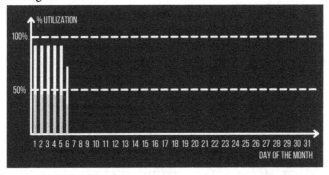

Figure 4.6 – Application usage with a peak at the start of the month

- **Applications-bound demand**: You have two bounded applications: one consumes the output of the other. This example is the most difficult to be spotted if you don't know your application map, but it's the most effective to be automated, as illustrated in the following screenshot:

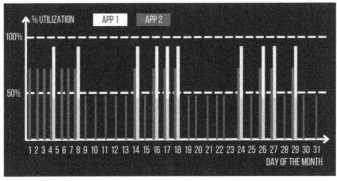

Figure 4.7 – Applications' usage/demand with bindings

- **Question 4**: *Can you easily engage the application developer teams for support?*

  Depending on the application and the level of support, if the application is off-the-shelf, you will likely have supplier support. If it is a custom application, you may have the developer team's support, and if it is an open source solution, you have forums and communities to ask other developers questions.

- **Question 5**: *What are your development/scripting skills (when using custom applications)?*

  C#, Java, Bash, Python, and so on.

Choosing from scheduler and orchestrator

To quickly implement the infrastructure changes when needed, you need to have a scheduler or, even better, an orchestrator able to run complex workflows, like the one we will build in the next section. This can be implemented in a variety of ways: using Azure native tools, such as Logic Apps and Functions, Azure Automation, or PowerShell scripting; by using third-party tools (Terraform, Chef, or any **infrastructure as code (IaC)** tool); or by using your own software tool.

> **Important Note:**
>
> There are many features to switch off (deallocate) VMs and other Azure services singularly. For example, you can use native automation to power off VMs, or even the new feature of **Start Virtual Machine (VM) on Connect** (`https://docs.microsoft.com/en-us/azure/virtual-desktop/start-virtual-machine-connect`). But what we often find is that an application relies on a set of different objects and services that need to start in a tangled way, which is why we'll learn about how to build an orchestrator for your application.

Let's start with understanding the scheduler part: you have a wide variety of programs that come bundled with the **operating system (OS)**, or open source or commercial software that allows you to run a script at a precise moment.

A **scheduler** allows you to basically run a procedure, and this procedure needs to contain all the workflow steps, data, and configurations to perform all the tasks in the correct order. Modifying the procedure means modifying the script and retesting it in all its steps, exactly like in a software development cycle.

> **Important Note:**
> Testing your scheduler is extremely important. Your scripts and workflows
> need to manage all the error types in the correct way to ensure the application
> is correctly stopped and restarted and that you won't be exposed to any kind of
> data loss.

If you're automating a small number of simple applications, you may use only a simple
scheduler, but if you're planning to automate many complex applications or if you plan to
extend the scheduling to an *on-demand* interface, you need to switch from a scheduler to
a proper orchestrator. The most frequent question I receive when I talk about schedulers
and orchestrators is: *Do I need to buy this third-party product because it supports Azure
with built-in integrations?* Well, 90% of the time, the answer is: *No, you don't need third-
party software, although it might ease the development.*

Every commercial software should have an internal developer life cycle provided by the
supplier, and some Azure objects may not be supported, either immediately or in the
future (this may be something your workload depends on), and you might end up with a
mix of supported and unsupported objects. This means implementing workarounds and
complicating the workflow and its maintenance. So, commercial software might not be the
answer to all your needs.

Building your own orchestrator

Since the beginning of this book, we have highlighted that you need to know your
infrastructure and your applications, and you need to know how to interact with the cloud
provider interfaces too (CLI commands, API, and so on). Since Microsoft brings you
PowerShell cmdlets, AZ CLI commands, and APIs, you could implement your workload
by yourself, keeping your integrations totally agnostic from any third-party product (and
their meta-languages).

If you're brave enough, you may start thinking you can create your own custom
orchestrator. It's totally true: if you're planning to use the orchestrator only for cloud cost
optimization, it's quite simple, and it's totally feasible. If you think about the most complex
cost-control/resource-shaping workflow, you'll end up realizing that its minimal, atomic
part is something like this:

```
{
    step_id : "<uniq id>",
    step_type : "<action name>",
    resource_id : "<resource id>",
    parameters : [],
```

```
    on_error : "<action on error>"
}
```

After identifying the step with a unique **identifier** (**ID**), we give it a *talking* name and assign the related resource ID. We also define what action will be performed in case of error.

Let's get real with an example, as follows:

```
{
    step_id : "431d4490",
    step_type : "poweroff_vm",
    resource_id : "/subscriptions/..........",
    parameters : [],
    on_error : "goto 081f1bd3455b"
}
```

If you build a **JavaScript Object Notation** (**JSON**) workflow descriptor file with atomic actions such as these, you may define an entire sequence of steps to reach a target. Then, your orchestrator should *only* read the descriptor file to get the sequence of all the single steps one by one, and then get the step type and engage the proper action on the resource (eventually, using the optional parameters).

The most basic orchestrator to stop/start your VM is offered as an embedded feature in Azure, named StartStopV2 (https://docs.microsoft.com/en-us/azure/azure-functions/start-stop-vms/overview). This feature relies on Azure Functions and Azure Logic Apps and will allow you to use a sequenced or scheduled pattern, and define your order of stop/start and the related tagging, along with providing valuable information through Application Insights.

But let's go back to building your own logic, regardless of the tool you are planning to use to implement it, which can be a mix of all the options we are learning about.

In high-level view, this design is a for loop on all the steps and has a lot of execution methods, one for each step type, with some error-management logic, as displayed in the following workflow diagram:

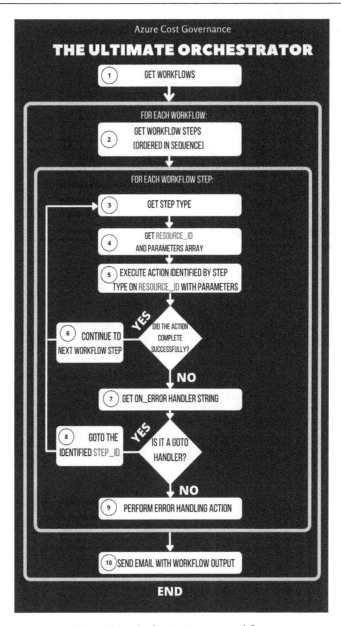

Figure 4.8 – Orchestrator core workflow

In this way, you've implemented the core portion of your orchestrator. If you couple it with a scheduler, you're more than halfway across, and you can start any complex workflow by simply defining a JSON descriptor. At this point, altering the workflow, step sequence, or creating a new automation workflow, will be as simple as editing the .json file.

Your world does not end in the cloud

Now that you have your orchestrator (either custom-built or third-party—it makes no difference at this point), you are planning to work on the following two main types of resources:

- **PaaS/SaaS resources**: This is where the cloud provider gives you everything you need to orchestrate the resource.

- **IaaS resources**: This is where the cloud provider gives you automation control only on the VMs, not on your services or applications installed in the VMs.

Let's brush up one more time on the MyCRM application from *Chapter 2, What Does Your Cloud Spending Look Like?*, where we learned about tagging. It was composed of the following:

- One SQL Server database in PaaS

- Two backend application servers installed on two VMs

- Two frontend application servers installed on two VMs

- One Azure firewall

- One frontend application gateway

- One backend load balancer

Let's suppose that, on weekends, the workload sinks and allows you to rescale the component to half of the SKU. Analyzing your application with the demand-shaping approach used in *Chapter 2, What Does Your Cloud Spending Look Like?*, it's clear that you need to start with the four VMs (the frontend and backend VMs of the following example) and the **PaaS SQL server**. You may start with an approach like the one in the following diagram, an MVP of the workflow:

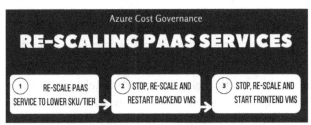

Figure 4.9 – Rescaling a SQL PaaS service (MVP)

So, this approach will work for sure, but if you must pay attention to the following points, then it's not the best approach:

- Minimize service downtime
- Minimize possible data loss or data corruption

The reason it's not the best approach is that you are trying to rescale an open database that may have queries running from the backend VMs, so when the SQL server then becomes available, you shut down the backend VMs, with dangling sessions on the frontend VMs, and so on. To avoid possible data corruption or data loss, you can move to a safer approach, as depicted here:

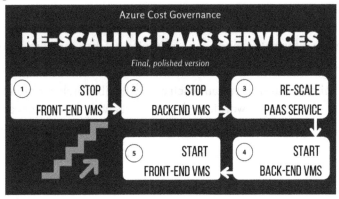

Figure 4.10 – Rescaling a PaaS service (final version)

This new version of the workflow will make sure no traffic will reach the backend and the database during all the PaaS service rescaling procedures.

> **Tip:**
>
> In the case of more complex applications, you may want to remove VMs from the load balancers or the application gateway balancing pools (this is called a **drain** procedure). This allows you to maintain the service with one system *chain* while you work on resizing the other *chain*.

We're assuming that you can start and stop the VMs and the MyCRM application, and that services start and stop correctly with their OS. Unfortunately, especially on older (legacy) applications, this is not always the case. There are applications that are composed of many services on a single node (which is a VM), and you have a start-and-stop sequence to bring it up or down successfully. To complicate our scenario a little more, suppose that 50% of the time, one crucial service does not start successfully, and you need to manage the error, correct the boundary conditions, and try to restart only that specific service.

Now, you surely start understanding the heading title: you cannot stop at what the cloud provider brings you. You need to dig deep until you get what your infrastructure or your application needs. This requires a little more development on the orchestrator to implement the following:

- Steps able to run commands and scripts in the VMs

- Steps able to check conditions (for example: *Is a specific Transmission Control Protocol (TCP) port open and listening? Have you found the word starting in a logfile?*)

Fortunately, Microsoft provides a way to do so, with the run-command subcommand of the AZ CLI. For more information and usage examples, refer to the official documentation at https://docs.microsoft.com/en-us/cli/azure/vm/run-command.

With those powerful commands, you can reach a very flexible orchestration boilerplate. You can assemble a very complex workflow that interacts with the services and is able to recover from an error condition. Here is a workflow diagram for our additions:

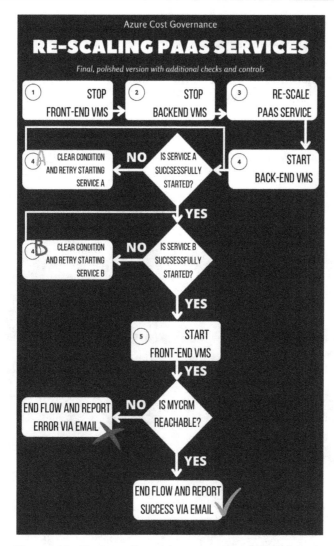

Figure 4.11 – PaaS rescaling orchestrator with additional checks

As you can imagine, we can add checks and actions on the VMs until we have a robust workflow that is able to bring the application resources up and down in a predictable way (let's assume that 98% of the starts are successful). A robust workflow should also check every possible error condition (that can be predicted) before starting any action that could fail, and after executing any further action, it should check for a successful output indicator (such as a log file line, a process in memory, an opened socket, and so on). If you plan to use Azure native tools, an Automation account with runbooks will help you keep track of everything that happens during the runtime of each automation job.

The same logic can be executed in many ways—for example, using Azure Automation and a runbook that will translate our generic workflow into your demand shaping script, or an Azure Logic App, such as the one in the next screenshot, where you can easily find pre-built connectors that will stop and start the chosen VM, and a link to an Automation account for the tier change of the PaaS service(s).

Every day at night, this Logic App will power off the frontend and the backend VM(s), then scale down the PaaS service (this can be Cosmos DB, Azure SQL, and so on) through an Automation job, then try starting the backend VM. It will then wait *until* the provisioning status of the backend VM is true (via the `jobdone` variable), then start the frontend VM, check with Application Insights whether the website is up, and send a confirmation or failure email depending on the outcome. Here is a workflow for the same:

Figure 4.12 – Our PaaS scaling workflow implemented with Logic Apps

Please remember that building a robust workflow may require time and a lot of testing (because you also need to simulate the error conditions, and you may need to develop ad hoc stubs to achieve this target).

In my humble opinion, you don't have to start implementing the most complex workflow with every condition. Let's keep it simple, start from some minimal unoptimized actions, and then add steps when you are confident with all the components (tools, command, service, credentials, and so on) and error handling.

On-demand re-scaling

In the *Understanding what you need to control* section, we saw some examples of workloads: some were on a scheduled basis (for example, working days versus weekends), but the last one depicts a *workload dependence* between two business applications. This is a common example of the **producer-consumer pattern** (see https://en.wikipedia. org/wiki/Producer%E2%80%93consumer_problem), where one application generates data/work used by another application. There are a lot of online examples of this pattern, but the important thing is that one application (the producer) could manage the rescaling of the other one (the consumer) based on its needs.

In the previous sections, we learned about orchestrators, and we started thinking about engaging the workflows with a scheduled approach. Here, we need to add a feature to our orchestrator: a way to start a workflow programmatically (and in a secure way, of course).

For example, we could create an API to start a given workflow and check the API key of the caller to ensure that the workflow engagement is legitimate.

In this case, you may prepare two workflows, as follows:

- One to rescale to higher SKUs
- One to rescale to lower SKUs

Then, modify the two applications to call the orchestrator, engage the correct action (and wait for the workflow completion), and then continue with normal operations.

The following workflow diagram shows you the orchestrator for the producer/consumer pattern:

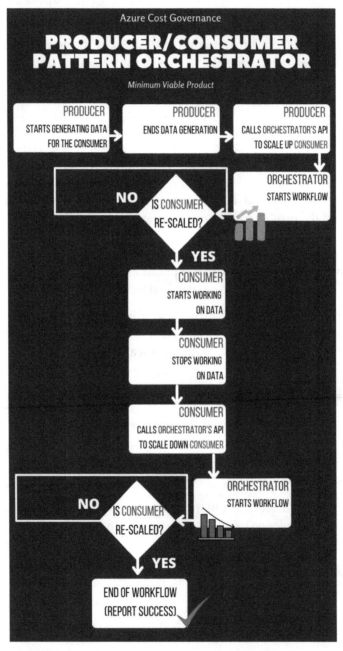

Figure 4.13 – Orchestrator for the producer/consumer pattern MVP

> **Tip:**
> Usually, rescaling to lower SKUs means that the application has stopped working on the produced data, so it may be the same application that asks the orchestrator to rescale itself down.

As usual, let's complicate our approach a little bit.

Currently, the producer can ask the orchestrator to provide more resources to the consumer when needed, but your needs may not be related only to cost saving—you may need the consumer to complete their job in a specific timeframe without overallocating resources.

Let's assume you may have three types of workloads for the consumer, as follows:

- **Workload A** is sequential, low-resource usage.
- **Workload B** involves a big, long run of parallel work.
- **Workload C** is sequential and very CPU-intensive.

In this case, you need to have four workflows with different SKUs for every component of the consumer (since you have different needs), as follows:

- One to rescale to SKUs according to workload A (it will use a low SKU, maximizing the saving)
- One to rescale to SKUs according to workload B (it will use expensive memory-oriented resources)
- One to rescale to SKUs according to workload C (it will use very expensive CPU-oriented resources)
- One to rescale to the lowest SKUs

Now that the orchestrator knows how to rescale objects (and you've extensively tested the workflows), you need to modify the producer to call the orchestrator and engage the correct workflow based on its specific needs in order to have the best consumer possible for that elaboration. The workflow becomes something like this:

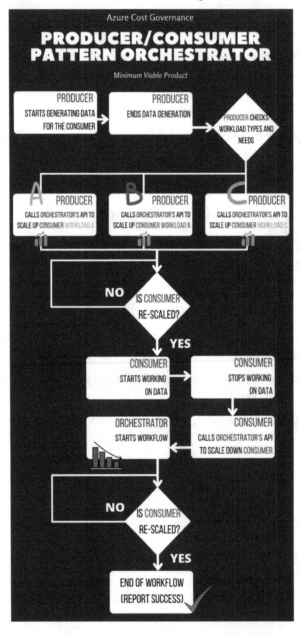

Figure 4.14 – Orchestrator for the producer/consumer pattern (final version)

At this point, you may start thinking that passing parameters and having an internal log for the orchestrator is a good and right thing to do, and you may be right: an orchestrator is a tool that can save you time and money, but as with every tool, it all depends on how it is used.

Let's return to the producer/consumer example. If the producer does not clearly understand the workload type, you may end up with the following:

- A lot of rescaling of workloads B or C and losing cost optimization
- A lot of rescaling of workload A, having delays in consuming data, and displeasing business users

Therefore, you need to carefully track who is engaging the workflow, when it's engaged, and which workflow is engaged, because sooner or later, you'll need to justify the infrastructure behavior and the performance versus the related costs.

This reasoning is overall true when dealing with VMs, but if you are using other Azure services, things definitely get easier, as outlined here:

- VM scale sets manage all this complexity for you. You can create metrics based on all the VM telemetry to create scale-in/scale-out flows.

- For PaaS services, most of them already scale in/scale out based on metrics configured by the user.

- Custom applications that use the producer/consumer pattern may take advantage of serverless code with Azure Functions (which has a durable capability) to trigger the autoscaling.

- For other services such as Service Fabric and **Azure Kubernetes Service** (**AKS**), the autoscaling is embedded in the orchestrator and will do the work for you according to your requirements and configuration.

The decisions you will need to make are only around how and when these services must scale, as they implicitly assume you are using the right tier and sizing and will only scale up and down according to traffic peak or performance issues. The same is valid for the next section, where we'll learn about horizontal scaling as a general rule, which, in many cases, is a concept already embedded in the PaaS service.

Focusing on other scale-up patterns

In the previous sections, we identified scale-up patterns that allow us to have the right computational resources when needed and shut down or scale down unused resources for cost-saving purposes.

But is vertical scaling the only way we can use our orchestrator and save money on the monthly bill? It surely is the simplest, but not the only one. In *Chapter 2, What Does Your Cloud Spending Look Like?*, we described vertical scaling and horizontal scaling and, depending on your application's architecture, it may be simpler to horizontally scale up the application during the day and scale it down during off-hours.

A very high-level simplification of a horizontal-scaling workload can be based on the already created workflow and ready-to-spin VMs. The precondition is that the VMs involved in this service are already behind a load balancer and are tagged with an identifying tag, as depicted here:

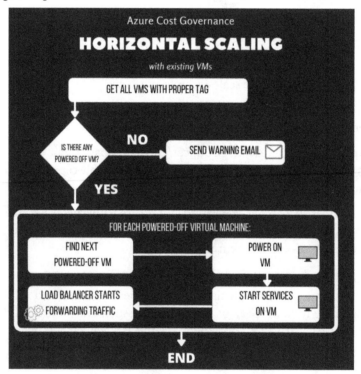

Figure 4.15 – Horizontal scaling with existing VMs

Of course, having a limited number of pre-created, ready-to-run VMs is not always applicable, but if we want to create new VMs automatically from an Azure image, the workflow only slightly changes, as we can see here:

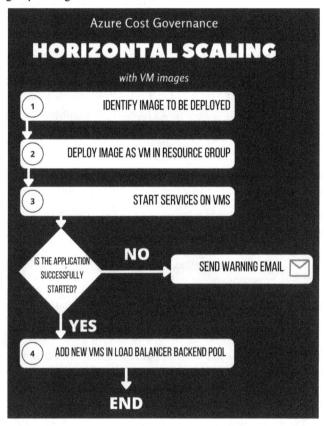

Figure 4.16 – Adding VM images to the horizontal scaling

Please remember that we're analyzing the horizontal scaling from an orchestrator point of view. If your need is only related to VMs, you can also rely, for example, on **Azure scale sets**.

Scale sets allow you to manage a load-balanced group of VMs that can automatically scale up or down depending on either a manual schedule or on the workload's demand. You can find additional information about Azure scale sets in the official documentation at `https://docs.microsoft.com/en-us/azure/virtual-machine-scale-sets/overview`.

Drawback in rescaling PaaS and SaaS

We've already faced the fact that rescaling IaaS components means shutting down the services and the VMs, with a business service downtime for the end user. You may be starting to think that IaaS is not so practical and that switching to PaaS or SaaS will relieve you from the downtime hardships; *the simple answer is: it depends.*

PaaS and SaaS are run on a classical infrastructure, managed by someone else, but the principles are similar: compute, storage, network, and memory, and adding memory or compute power very often requires a *restart* of the resources where your PaaS service resides. Backend VMs need to be restarted, disks need to be changed from one tier to another, and containers need to be redeployed with a new configuration.

Of course, PaaS and SaaS are more flexible than IaaS and they relieve you from having to manage every single configuration in detail since, on Azure, those services have well-defined blocked tiers, and every service comes with a powerful API or cmdlet to be managed.

Changing an Azure SQL database tier can be as simple as running this command (using **Azure-CLI version 2.30.0**):

```
az sql db update -g <ResGroupName> -s
<ServerName> -n <DatabaseName> --edition
(Basic|Standard|Premium|GeneralPurpose|BusinessCritical|
Hyperscale) --service-objective <TierName>
```

The previous AZ CLI command has the default parameters for resizing your SQL database. An example of this is provided here:

```
az sql db update -g ne-rg-MyCRM -s MyCRMsqlProd01 -n MyCRMdb
--edition Standard --service-objective S1
```

Switching from one tier to another is far simpler than running the same server in a VM group: all we need is one API call or one command with the right parameter, and that's it. But remember that in some cases, you'll have a little downtime, and your applications need to be aware that they may lose the connection to the modified resource. Also, they need to re-establish or refresh the connection as needed, to be fully functional when the service becomes available in the new tier.

> **Tip:**
> Keep in mind that services can be unreachable or become unresponsive even in the cloud. This kind of application resilience to PaaS or SaaS services is surely useful for cost saving, but it's very important in *any aspect* of your infrastructure and application design because it ensures no manual intervention for *reducing the business service downtime to a minimum* for your users.

Typical per-resource-type cost-saving approach

In the previous sections, we mainly concentrated on the IaaS optimizations (up/down and vertical/horizontal scaling), but please keep in mind what we learned in *Chapter 2, What Does Your Cloud Spending Look Like?*, under the *The benefits of PaaS services—demand shaping* section, where we discussed some strategies to optimize the PaaS/SaaS resource SKUs to your demand with the help of some examples, such as **Just-in-Time (JIT) provisioning**.

Using those hints, you need to analyze every PaaS/SaaS service used by your applications to create parametrized methods for your orchestrator to operate on the different resource types. You will create a *library* of those methods operating on specific services, allowing you to turn the right screw at the right time, as our experienced *cloud watchmaker*.

Azure offers many tools, and features are added constantly to assess and respond to the performance versus sizing conversation, such as Azure Advisor or **Data Migration Assistant** (**DMA**) and Azure Migrate. These tools should be used as a best practice for any wise migration project but can be also used to assess your cloud infrastructure (especially the **Azure Advisor** page) continuously and periodically.

JIT provisioning

Enforcing your cost governance policy should be grounded on the concept of **JIT provisioning**, leaving no space for wasted resources. Many cloud teams use an IAC tool, either native, with Azure Automation and **Azure Resource Manager** (**ARM**) templates, or tools such as **Bicep** (`https://docs.microsoft.com/en-us/azure/azure-resource-manager/bicep/overview`) and **Terraform**. These tools require a good level of cloud skills to create the initial configuration, but have clear benefits that made them very popular with cloud teams, such as the following:

- Offloading laborious and interminable tasks
- Avoiding human error
- Allowing for a higher scale of resources and users

The importance of JIT is also in the resource decommissioning part: as we happily jump on creating Azure objects for any new user/project, we should always plan to deallocate such resources when they are no longer needed. The beauty of it is that recreating resources at this point is just one click away, so our internal infrastructure users should not be scared of our decommissioning policies.

Summary

This chapter helped us start painting a picture of cost-saving techniques that you can use to both lower your Azure costs and keep them down. We learned about the principles of right-sizing and how to deal with cloud resources with the correct mindset. Then, we explored a set of examples and workflows that can be easily applied to your Azure environment to assess, automate, and improve your costs, such as gradually building and enriching your own orchestrator that can be invoked from your application and, in fact, perform on-demand scaling and many other savings.

The next chapter will be dedicated to cleanup, which means recognizing and dealing with resources that are completely unutilized or orphaned yet paid at the end of the month.

Questions

1. Where can you get a list of VMs to be right-sized?

2. Which types of managed disk storage can you choose from (cheapest to most expensive)?

3. Can you downsize a VM with no downtime/impact?

Further reading

- Migration best practices for costs: `https://docs.microsoft.com/en-us/azure/cloud-adoption-framework/migrate/azure-best-practices/migrate-best-practices-costs`

- On-resources throttling:

  - `https://docs.microsoft.com/en-us/azure/azure-resource-manager/management/request-limits-and-throttling`

  - `https://docs.microsoft.com/en-us/azure/service-bus-messaging/service-bus-throttling`

  - `https://docs.microsoft.com/en-us/troubleshoot/azure/virtual-machines/troubleshooting-throttling-errors`

5
Planning for Cost Savings – Cleanup

Now that we've learned how to analyze, process, and implement right-sizing techniques, the next part of the cost-saving section is dedicated to cleaning up. This means having full control of your cloud infrastructure and being able to access, quarantine, and delete unused resources where possible.

In this chapter, we will find and remove services and configurations that are deemed unnecessary, either manually or via automation. We will also recognize all the resources that were used temporarily for migration and may have been forgotten. We'll also learn how to analyze cost spikes that may be the result of unattached resources, shared resources, and even unused subscriptions, along with their security implications.

In this chapter, we will cover the following topics:

- Cleaning up the cloud resources
- Byproducts of migrating to Azure
- Identifying cost spikes

By completing this chapter, you will have a clear understanding of the most used best practices and techniques for identifying, managing, and containing the cost of unused or ghost resources.

Technical requirements

For this chapter, you'll need the following:

- A computer with internet connectivity.

- Access to the Azure portal.

- Privileges to access Cost Management information (see `https://docs.`
  `microsoft.com/en-us/azure/cost-management-billing/costs/`
  `assign-access-acm-data`).

- To execute cost report scripts, you'll need a host with at least **PHP 7.4** and a **MySQL**
  or **MariaDB** database. It's also mandatory to have the initialized and imported
  billing data from *Chapter 3, Monitoring Costs*.

You can find the code for this chapter here: `https://github.com/`
`PacktPublishing/The-Road-to-Azure-Cost-Governance/tree/main/`
`Chapter%205`.

Cleaning up the cloud resources

Another important part of cloud cost saving is the cleanup operation. When dealing with
a large number of workloads or complex projects, lots of resources are created just as
a transitional step and are often forgotten about and paid for. In this section, we'll learn
how to identify and clean up all the unused and unattached resources in your virtual
data center.

Cleaning up unused items, as a first approach, represents one of the short-term, quick-
win techniques for cost-saving. But, on the other hand, if they're inserted into a recurring
process, this will help you uncover any unassigned or unutilized infrastructure (with
operational downfalls) and, in general, uncover the gaps in your operational processes
that might need extra effort and have a wider impact than costs. In addition, you should
plan to periodically assess the evolution of your infrastructure for any ghost resources that
may have been left unassigned and unused.

Free stuff on Azure

Before we dive into cleaning up, let's make sure we know that there are lots of free resources in Azure that let you *play* with the cloud at no cost. At the time of writing, there are two types of free Azure resources:

- **Services that are always free, no matter the size, contract, or subscription types**: Azure Advisor, **Azure Kubernetes Service** (**AKS** – only the orchestrator part), Azure Batch, Azure Policy, Data Catalog, DevTest Labs, inbound inter-VNet data transfer, public Load Balancer, Security Center assessments and recommendations, Service Fabric (the orchestrator part), SQL Server 2019 Developer Edition, and Visual Studio Code.

- **Services that are free up to a specific usage/tier**: Anomaly Detector, App Bot Service, App Configuration, App Service, Archive Storage, Automation, Azure AD, Azure Cosmos DB, Azure DB for MySQL, Azure DB for PostgreSQL, Azure DevOps, Azure SignalR, Bandwidth (data transfer), Blob Storage, Cognitive Search, Computer Vision, Container Registry, Content Moderator, Custom Vision, Data Factory, Event Grid, Face, File Storage, Form Recognizer, Functions, Key Vault, LUIS, Logic Apps, Machine Learning, Managed Disks, Media Services encoding, Network Watcher, Notification Hubs, Personalizer, QnA Maker, Service Bus, Spatial Anchors, SQL Database, Text Analytics, Translator, Linux VM B1s, Windows VM B1s, and Virtual Networks (VNets).

Detailed information on free Azure resources can be found here: `https://azure.microsoft.com/en-us/free`. However, we encourage you to take advantage of these free services, especially when cleaning up resources, which you can do by downgrading to a free tier; you'll still be able to *play* with them.

> **Important Note:**
>
> In addition to the free services and tiers, please bear in mind that all the features in preview are typically at lower or no cost, even if they can't be used for production purposes. This is particularly important when a feature is moving toward General Availability because its price will change, and you will get the final charges in your monthly bill.

Azure Resource Graph

Another important service to learn about, that might be helpful not just across this chapter but in your overall cloud journey, is **Azure Resource Graph**. This is a service that allows you to execute powerful queries (there is even the Resource Graph Explorer in the portal to help you with it) and scan your subscription(s) for resources, policies, and configuration. But for our exercise, to get all the changes in our resources (as explained at `https://docs.microsoft.com/en-us/azure/governance/resource-graph/how-to/get-resource-changes`) and update our **Configuration Management Database** (**CMDB**) and any compliance tool, we'll need this type of information.

A useful tool to get a complete governance report our of your subscription(s) is the Azure Governance Visualizer, which can be found here: `https://github.com/JulianHayward/Azure-MG-Sub-Governance-Reporting`.

This tool will help you build a visual hierarchy of your Azure resources and can provide an aggregate consumption report across all the analyzed scopes.

Unassociated services

In your virtual data center, from time to time, you may find a bunch of services that are unassociated with other resources or orphaned. Typically, these are not in use, and they are the result of someone forgetting to delete the complete set of resources, or developers' teams creating a proof of concept (POC) or sample and forgetting to clean up their environment.

An Azure simple single **virtual machine** (**VM**), for example, when created via the Azure portal or CLI, will spawn a bunch of collateral resources that can be free of charge such as a **VNet** or paid for (such as a **public static IP address**). The key takeaway of this section is that every cloud service that does not have a specific purpose must be tracked, analyzed, and, if deemed unnecessary, disposed of.

The most common resources that are left orphaned in Azure are as follows:

- **Unattached disks**: These resources represent disks that are not attached to any VM. They might be kept as a backup for deallocated VMs, or they might be simply the result of a VM deletion that did not include their storage. In any case, for unused storage disks, you will need to analyze, validate, and delete or temporarily move them to a cheaper tier.

- **Unused storage accounts**: As for disks, you might find storage accounts with a few files that have not been accessed for months. You will need to notify the owner that these accounts are not being used and that they will be removed at some point in your cleanup process.

- **Static IP addresses**: We are not talking about big numbers, but allocating unused or unattached resources is not right. This is valid for static IP addresses as well: these static addresses once gave you access to a VM that is no longer allocated or a service that is no longer needed.

> **Important Note:**
> Using a public IPv4 address for unused resources has additional implications. There is a chronic lack of IP addresses that sometimes makes these resources more valuable and expensive, depending on the region. This concept applies throughout the whole set of cloud objects, where prices change (albeit slightly) region by region and according to many local variables.

- **Logs**: This is a tricky one. When you set up an Azure Monitor solution, you probably want to test and try all the possible logging features. But logs are charged by the total storage retention and transaction (ingestion). Chances are, after some time, you'll find yourself with a large monthly bill for several useless logs that you can get rid of, as explained in this Microsoft article: `https://docs.microsoft.com/en-us/azure/azure-monitor/logs/manage-cost-storage`.

- **Snapshots and backups**: Similar to log retention, snapshots and backups, in general, are pieces of storage (and spending) that must become useless at some point because you have a newer version, or because the application changed. In any case, controlling the retention time of your snapshots is a good cost management practice, along with a good log rotation policy that will have older data overwritten instead of growing forever.

- **Hidden networking charges**: When setting up your virtual data center, you might have been tempted to replicate a networking space that's very similar to what you had on-premises. This led to several peering configurations that are probably unnecessary but still consume resources and money.

Now, let's look at a practical example of how to identify unused resources.

A practical example of unattached disks

You want to make sure you don't have any unattached disks in your Azure virtual data center. This can easily be done by searching through the portal for any managed disk that has a - (*dash*) as the owner, as explained in the following documentation: `https://docs.microsoft.com/en-us/azure/virtual-machines/disks-find-unattached-portal`. The following screenshot shows the list of unattached managed disks:

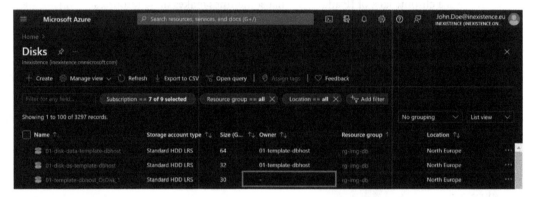

Figure 5.1 – Unattached managed disks

The same list can be obtained via the Azure CLI and exported in a `.tsv` (tab separated) file:

```
az disk list --query '[?managedBy==`null`].[id]' -o tsv
```

Once you have found a disk that matches this requirement, by clicking on it, the page will display a status of **Disk State=Unattached**. This means that no VM is linked to that disk, and you will have to define a workflow for hibernating and/or canceling that disk(s), according to your IT department's policy.

> **Important Note:**
> For classic (unmanaged) disks, the search will show the **Attached to** column, similar to the managed disk view: if no resource is listed, the disk is orphaned.

We have prepared a diagram to summarize the workflow of deleting unattached storage:

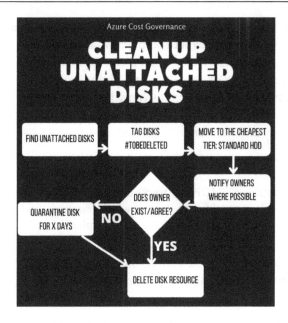

Figure 5.2 – Workflow of hibernating and/or removing unattached disks

Of course, you will have to define your internal deletion policy to allow a grace period before any relevant information gets deleted.

> **Tip:**
>
> Azure managed disks are charged by their target size, regardless of their effective usage. For example, if you need 500 GiB, you will pay for 512 GiB, which is the closest disk size (a P20 for premium disks).
>
> This also means that if you allocate even 1 GiB above the predefined size, you are going to pay for the next size: so, 513 GiB will be charged as a 1,024 GiB (P30). I have seen these errors due to migration tools and automated sizing, so I recommend searching through your managed disk sizing for any *suspicious* size that might incur unnecessary costs and cleaning them up by reverting to the originally intended size as soon as possible.

Automated waste disposal

A very important thing to always keep in mind is that, except for the free services mentioned earlier, nearly everything you create on a cloud environment has a cost, and if you forget to delete dismissed or unused resources, you may incur unjustified extra costs. For this reason, it's crucial to ensure that all unnecessary items are deleted, possibly according to a retention/contingency policy.

In *Chapter 2, What Does Your Cloud Spending Look Like?*, we learned about tagging. The same technique can be used to ensure that you track what you need to delete and when. For example, when you dismiss a resource or an entire application, you could add a tag such as `delete_after : "YYYY-MM-DD"` and periodically check (maybe once a week) with Azure Automation or with our custom scheduler, as we learned in *Chapter 4, Planning for Cost Savings – Right Sizing*, every resource with that tag to identify which resources need to be safely and permanently deleted.

> **Tip:**
> Having a well maintained CMDB that tracks every resource for each business application is crucial for finding out which resources can be tagged, quarantined, and then permanently deleted once dismissed.

Unfortunately, some automatically generated resources may slip from the tagging procedure (since they are generated directly by the Azure backend). Some examples are as follows:

- Backups
- Snapshots
- Public IP addresses
- Network Watcher (auto-enabled)

These resources may be hidden for months before their total monthly spending becomes *noticeable*: for example, each snapshot can be very cheap, but if you keep piling up snapshots, the total cost can rise to a large amount at the end of the month. You must use the **Meter Category/Subcategory analysis** to identify such hideous kinds of costs and implement automation to delete useless snapshots or backups after a retention period, or even implement a process to monitor those costs and guard them.

Now, returning to the automation task, you can create a simple scheduled script that extracts all the cloud resources, checks the tag's value, automatically deletes the resource, and then sends a notification. The following workflow diagram illustrates this process (recall the tagging suggestions from *Chapter 2, What Does Your Cloud Spending Look Like?*):

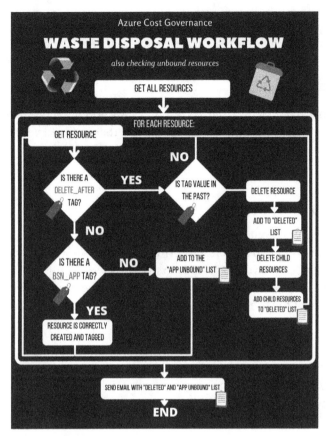

Figure 5.3 – Waste disposal with unbound resources checks workflow

Please note that the previous workflow relies on tagging best practices to do the following:

- Make sure all the relevant resources are tagged and cleaned up.

- Identify which resource should be consciously deleted (based on the delete_ after tag) and autonomously delete it.

- Identify the resources unrelated to a business application and notify them because of the following reasons:

  - These resources could be hidden costs that are automatically generated (for example, backup snapshots, NICs, and so on).

  - They could be manually created, such as temporary resources that are not tagged for any reason (for example, disks).

  - They could be legit resources and we simply forgot to add the **BSN_APP** tag (as in the workflow), and we should be aware of that so that we can remediate it.

> **Tip:**
> It's very useful to include the costs of these resources (from their creation, or from the selected month to the current date) in the mail report, just to monitor the extra costs and savings from the automation process.

Automating the process of deleting unused resources is a good practice. Now, let's dive into the unutilized, yet allocated, networking resources that may add very little to the monthly bill but have several additional implications that require them to be cleaned up properly.

Cleaning up networking resources

In my experience, many customers approach cloud networking as they would their on-premises set of switches and routers. Unfortunately, this is not the best approach – not from a configuration perspective – as it leads to many issues and misconfigurations, and especially not from a cost perspective. In your data center, the network cabling, switching, and routing are a part of the infrastructure's initial cost, but in the cloud, this is paid for by consumption, just like any other service.

In public clouds, bandwidth is charged and several things will add to your billing, even if they are probably unused. As explained at https://azure.microsoft.com/en-us/pricing/details/virtual-network/, when you want to connect the network between two VNets, you should use a peering configuration, which is a paid service.

In the hub and spoke topologies, which are commonly used in enterprises and large companies, peering is used to connect the spokes to the main hub, and the traffic between the hub and spokes generates VNet charges. During the design phase, where possible, you should keep your top talking applications in the same VNet. Once your migration/refactoring/delivery project is complete, we recommend checking that all the remaining peerings are necessary and removing those that aren't.

In this section, we have learned, by using practical examples from our experience in the field, how to get rid of unattached and unutilized resources, from managed disks that no longer have a related VM to any other possible ghost resource that is spending money without providing real value. Now, let's look at what happens when you migrate workloads using tools, processes, and a methodology that sometimes leaves resources behind, and how to properly address them.

Byproducts of migrating to Azure

When you approach a large resource migration project (for example, from an on-premises data center) or an application upgrade, it's normal to have resource duplication, parallel environments, large database backups, or dumps around your infrastructure.

Azure Migrate and Azure Site Recovery

When you have to migrate applications and services to Azure, there are great tools that can help you define your landing resources sizes and tiers.

Azure Migrate is a hub that lets you discover, assess, and migrate your on-premises applications, providing you with insights into how your application will perform in the cloud, as well as the associated costs. It includes several integrated tools and even allows for additional third-party/ISV integrations. For our cleanup section, we will look at Dependency Analysis, which can be very useful for understanding if you can get rid of resources that are unused or under-utilized without impacting other applications. For more information, please refer to `https://docs.microsoft.com/en-us/azure/migrate/`.

Azure Site Recovery is a **Business Continuity and Disaster Recovery (BCDR)** strategy tool that can manage VM replication from on-premises virtual and physical servers between Azure regions and even AWS VMs. Please refer to `https://docs.microsoft.com/en-us/azure/site-recovery/site-recovery-overview` for this.

Using native tools for migration is typically the best practice as you will be able to safely automate the process and get the most out of the applications' performance and cost information. Please be aware that not all workloads are supported and, in some cases, you might have to use third-party migration tools that can make your migration a bit more complex. Let's look at some common practices that might help you either way.

Migration tagging strategy

It's very important to have a structured approach and to keep track of this resource overallocation due to duplication or data protection. My recommendation is to add a tag specifying that some resources are part of the migration project and need to be checked and decommissioned within a specific date or month. In this way, after the end of the migration project, you just have to query resources by tag and you'll easily be able to identify what you need to double-check, power off, and safely decommission, as shown in the following screenshot, by using the highlighted filtering options:

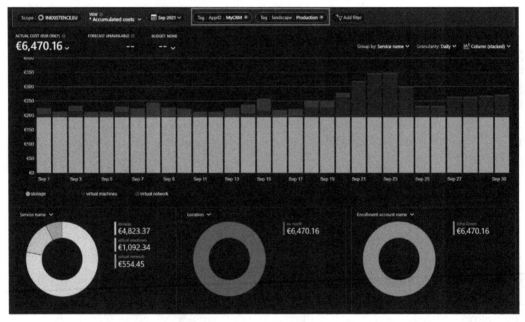

Figure 5.4 – Cost management (filtering by tag)

Of course, all the examples shown in the previous section about inducted costs and derived optimizations (such as disks) should be applied.

If you implement a correct tagging policy for high-impact projects, I also suggest that you verify and monitor the resources' costs using the Azure Cost Management tool, filtering only by tag. This will highlight any hidden resources or costs.

Now, let's dig deeper into this concept with a practical example that will explain how tagging can solve most of the issues that are created when migrating using temporary resources.

A practical example of migrating byproduct tagging

Let's say you are planning an operating system upgrade because of the end-of-support and end-of-life security problems in your cloud infrastructure.

Let's consider your three environments:

- Development
- Testing
- Production

For each environment, starting from the lower-impact one with an application-by-application approach, the steps for the cleanup process are displayed in the following diagram. At the end of the project, you'll have a clear extraction by being able to tag all the resources you need to delete, and you will be able to easily automate how unused resources are disposed of.

The following diagram is a draft workflow of a high-level approach to a massive upgrade project, where you will have to check if each resource must be upgraded, then proceed with the proper tagging, including for temporary resources:

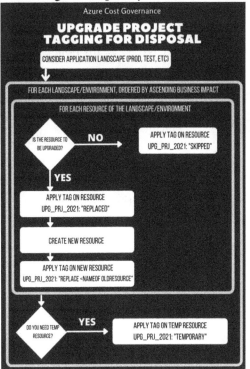

Figure 5.5 – High-level approach to a massive upgrade project

This will allow you to control what is left out of the upgrade and, most of all, the temporary resources that, after a while, can be promptly deleted.

> **Tip:**
> If compatible with your project's constraints, please consider creating the parallel environments one at a time. For example, when you switch the application to the *new development environment*, you'll power off the *old development environment*, start creating the *new test environment*, and so on. This approach will limit the co-existence of resources, thus reducing costs.

Now that we've learned how to get rid of unutilized migrations' byproducts, let's look at a cloud best practice that brings additional clarity and discipline to the resources' cleanup.

The snooze and stop workflow

A common practice for large companies, especially when migrating many VMs with unclear scopes and owners, is to configure and enforce a **snooze and stop policy**. The concept behind this is that in on-premises environments, users sometimes provision a server and then forget about it. This server can then be migrated to the cloud for continuity, even if the user is unreachable.

This typically prompts for a clear and shared company policy that will inform your internal users of the following:

- A VM that has a 5% CPU usage for more than 30 days will be snoozed (switched off) and its disks will be switched to the lowest tier possible (for example, standard HDD).
- All the impacted resources should be immediately tagged as snoozed to avoid accidental power on.
- If nobody claims that particular VM within another 30 days, it will be deleted (keeping the storage at the minimum tier; that is, standard HDD).
- After another 30 days, the associated storage will be deleted as well.

You can use native tools such as VM Insight and Azure Automation, or even Logic Apps, to get all the relevant information and implement this logic. Just be aware that some tools that rely on Log Analytics might result in additional charges for logging the relevant information.

Here is a workflow diagram of the snooze and stop process:

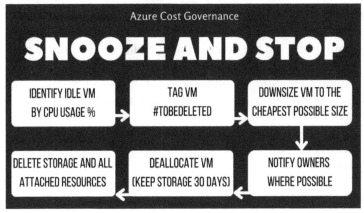

Figure 5.6 – Snooze and stop workflow (use with caution!)

This policy can save you a lot of money in the long run and not just on compute and storage costs, but also the operational cost of keeping workloads that are no longer used or claimed alive and maintaining them.

> **Important Note:**
>
> Make sure that the object/resource you are snoozing is highly advertised for deletion by sending notices via a snooze bulletin to all your internal customers. In my experience, there are many forgotten VMs, especially after migrations, but it may still be possible that you are deleting an object that has its own use. If you are unsure of this, proceed with the snooze process and increase the storage retention to up to 6 months or even 1 year.

In this section, we learned how to identify, analyze, and contain costs that are the result of migrations. Now, let's identify cost spikes (and subsequently remove the unused resources where possible) by starting with cost reports.

Dealing with a CMDB

If you are approaching the cloud with cloud-native applications and no legacy stuff, you are in an optimal position to be able to use all the native services that allow you to safely manage your configurations. But the overwhelming reality of IT infrastructures is still tied to on-premises and legacy applications, which make a CMDB a necessary tool, especially when you're migrating to or adding cloud resources. Deleting a CMDB can be different, depending on your company's requirements, but it should include the following for every registered item:

- A unique identifier
- Name
- Description
- Owner
- Priority

In on-premises and CAPEX financial cost model environments, CMDB is a useful tool for tracking versions of hardware and all the implications of its obsolescence and amortization. So, it's natural that many companies coming from custom data centers have kept this strategy also with cloud services.

Updating a CMDB for legacy data centers might be easy since the pace of changes was bound to the hardware and software contracts in place. However, on the cloud, where a resource might live only for a few minutes, things get tricky.

Many options and vendors now offer public cloud integrated CMDB tools, while the option of integrating a custom tool is still valid thanks to the Azure APIs, which expose every cloud object that you need to map. This is the first step to building your company's application map, where you are registering all your cloud resources according to their business usage and priority and will be able to track application dependencies and, for the sake of this chapter, orphaned objects.

As you may recall from the *Tagging* section in *Chapter 2, What Does Your Cloud Spending Look Like?*, where we learned about the importance of having a good tagging process and policy, if your Azure environment is properly tagged, automatically populating a CMDB should be a piece of cake.

Identifying cost spikes

As we learned in *Chapter 1*, *Understanding Cloud Bills*, and *Chapter 3*, *Monitoring Costs*, at this stage, you should be familiar with all the tools that allow you to get a full picture of your spending and keep it predictable and under budget. Sometimes, however, there are specific conditions that will create unexpected spikes in your costs and will call for a prompt cleanup action. My recommendation is to *check the Azure Cost Management page daily* and look at a daily stacked view of your spending. The following screenshot shows the cost management page:

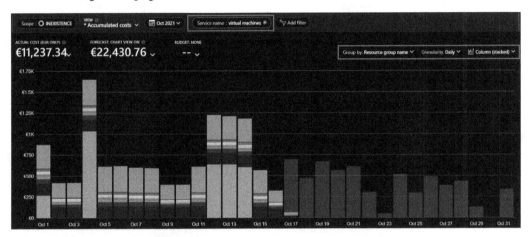

Figure 5.7 – Example of a daily stacked chart on your cost management page

In normal operations, you should be able to identify a nice clean pattern (just as the one pictured in the previous example) and immediately spot any spikes that do not resemble your normal pattern of operations.

Spikes can be caused by many underlying issues:

- An application that is calling a service too many times due to a bug

- A problem with the network prompting too many retries

- An application having an issue and logging *10 times* the amount of normal logs

- A spike in network egress traffic, leading to a security breach

- A serverless PaaS project suddenly having a spike in user requests

- A seemingly innocent change that was made in a database configuration that suddenly logged tons of storage

- A service under attack, such as from a DDoS attack, that will cause spikes in the egress traffic and transactions

> **Important Note:**
> When you buy reservations, the total amount that's purchased for that service will spike up on the day you reserved it. For it to be spread throughout the timeframe, you must use the **Amortized cost** option, where you'll be able to see the daily cost of the reservation operation, as we'll see in the next chapter.

If you set up your budgets and alerts correctly, as you learned in *Chapter 3, Monitoring Costs*, you should be able to spot any spikes within 1-2 days (which is the aggregation of the average time for Azure Billing).

> **Important Note:**
> Checking for these spikes can also uncover application issues and is considered a best practice for all cloud deployments.

But for all the unmonitored resources and workloads, my recommendation is to *look at your Azure spending daily* so that you can act quickly and fix any underlying issues that caused the spike.

> **Important Note:**
> There are several proactive services you can use to mitigate a security attack, so there will be cost implications of a DDoS attack.

Now, let's analyze a few cases where identifying spikes is complex and sometimes has collateral implications on spending resources.

Per-meter category split cost report

Setting budgets on costs or forecasts for each resource group, each application, or at the subscription level can help you identify spikes, but sometimes, this method has some drawbacks, as follows:

- A lot of manually defined budget alerts, which you need to maintain when adding resources.

- Resource group budget alerts, which may include resources that are tied to different applications.

- Budget alerts created at the subscription level may be too generic.

Budget alerts are a good start but require deep manual analysis to identify which resource is the extra-spending one. Previously, we learned how to automate cost details extraction using the Cost Management feature and ingest the costs into a custom database. If you implement daily extraction, coupled with daily ingestion, you may create a customized day-over-day mail report.

> **Important Note:**
> Daily cost exports do not have complete cost information for *today* (since we're still in the day). So, after the ingestion process, please consider working up to the full completed day: *yesterday*.

Once you have all the data, I suggest creating the following reports (with a tabular output sample):

- **Meter category day-versus-day**: This report compares costs split by Meter Category of *day-1* (yesterday) to *day-2* and highlights every spike after a certain threshold:

Azure Cost Governance
METER CATEGORY DAY-VERSUS-DAY REPORT

Meter Category	Day-2 (21-09-13)	Yesterday (21-09-14)	Delta cost	Delta %	TREND
STORAGE	132.98€	133.12€	0,14	0,10%	⋀
BANDWIDTH	32.25€	31.10€	-1,15	-3,69%	⋁
...					

Table 5.1 – The meter category day-versus-day report

You can find the code for this report in the `report_metercat_day_day.php` file on GitHub at `https://github.com/PacktPublishing/The-Road-to-Azure-Cost-Governance/blob/main/Chapter%205/report_metercat_day_day.php`.

- **Meter category current month-to-date versus (month-1)-to-date**: This report compares costs split by Meter category based on month-to-date (of this month) versus the month-to-date (of the previous month) costs:

	Azure Cost Governance **METER CATEGORY MONTH TO DATE COMPARISON REPORT**				
Meter Category	**August 1 to 13**	**September 1 to 13**	**Delta cost**	**Delta %**	**TREND**
STORAGE	132.98€	133.12€	0,14	0,10%	⋀
BANDWIDTH	32.25€	31.10€	-1,15	-3,69%	⋁
...					

Table 5.2 – The meter category month to date comparison report

You can find the code for this report in the `report_metercat_month_to_date.php` file on GitHub at `https://github.com/PacktPublishing/The-Road-to-Azure-Cost-Governance/blob/main/Chapter%205/report_metercat_month_to_date.php`.

- **Resource group day-versus-day**: This report compares costs split by resource group of *day-1* (yesterday) to *day-2* and highlights every spike after a certain percentage threshold:

	Azure Cost Governance **RESOURCE GROUP DAY-VERSUS-DAY REPORT**				
Res Group	**Day-2 (21-09-13)**	**Yesterday (21-09-14)**	**Delta cost**	**Delta %**	**TREND**
My-CRM-rg	132.98€	133.12€	0,14	0,10%	⋀
Infra-rg	32.25€	31.10€	-1,15	-3,69%	⋁
...					

Table 5.3 – The resource group day-versus-day report

You can find the code for this report in the `report_resourcegroup_day_day.php` file on GitHub at `https://github.com/PacktPublishing/The-Road-to-Azure-Cost-Governance/blob/main/Chapter%205/report_resourcegroup_day_day.php`.

- **Resource group current month-to-date versus (month-1)-to-date**: This report compares costs split by resource group based on month-to-date (this month) versus the month-to-date (previous month) costs:

	Azure Cost Governance				
	RESOURCE GROUP MONTH-TO-DATE REPORT				
Res Group	AUGUST	SEPTEMBER	Delta cost	Delta %	TREND
My-CRM-rg	132.98€	133.12€	0,14	0,10%	⋀
Infra-rg	32.25€	31.10€	-1,15	-3,69%	⋁
...					

Table 5.4 – The resource group month-to-date report

You can find the code for this report in the `report_resourcegroup_month_to_date.php` file on GitHub at `https://github.com/PacktPublishing/The-Road-to-Azure-Cost-Governance/blob/main/Chapter%205/report_resourcegroup_month_to_date.php`.

- **Per application day-versus-day**: This report compares costs split by application (using the *application ID* tag) of *day-1* (yesterday) to *day-2* and highlights every spike after a certain percentage threshold:

	Azure Cost Governance				
	APPLICATION DAY-VS-DAY REPORT				
APPLICATION	2021-09-13	2021-09-14	Delta cost	Delta %	TREND
MY CRM	132.98€	133.12€	0,14	0,10%	⋀
domain controller	32.25€	31.10€	-1,15	-3,69%	⋁
...					

Table 5.5 – The application day-versus-day report

You can find the code for this report in the `report_businessapp_day_day.php` file on GitHub at `https://github.com/PacktPublishing/The-Road-to-Azure-Cost-Governance/blob/main/Chapter%205/report_businessapp_day_day.php`.

- **Per application current month-to-date versus (month-1)-to-date**: This report compares costs split by application (using the *application ID* tag) based on month-to-date (this month) versus the month-to-date (previous month) costs:

Azure Cost Governance

APPLICATION MONTH-TO-DATE REPORT

APPLICATION	August 1-13	September 1-13	Delta cost	Delta %	TREND
MY CRM	132.98€	133.12€	0,14	0,10%	⋀
domain controller	32.25€	31.10€	-1,15	-3,69%	⋁
...					

Table 5.6 – The application month-to-date report

You can find the code for this report in the `report_businessapp_month_to_date.php` file on GitHub at `https://github.com/PacktPublishing/The-Road-to-Azure-Cost-Governance/blob/main/Chapter%205/report_businessapp_month_to_date.php`.

If you were able to deploy the script in *Chapter 3, Monitoring Costs*, in the *Thinking about an architecture for a custom cost management tool* section, then you will be able to follow up by adding these reports. You can find them here: `https://github.com/PacktPublishing/The-Road-to-Azure-Cost-Governance/tree/main/Chapter%205`.

In this section, we learned how to identify cost spikes by comparing cost reports. This practice should be included in a healthy recurrent cost governance process and be performed at least once per quarter. *But what about identifying hidden costs among shared resources?* We will cover this in the next section.

Shared resources

In the previous sections, we learned that the granularity of the representation mostly depends on the tagging completeness: without proper tagging or a fully populated CMDB, it's impossible to determine the costs of every application.

This was a necessary approximation to introduce some major concepts about cost control and its various representations.

The more you understand how to identify hidden costs and spikes, represent them, and keep them under control, the more you start to realize that they are mostly related to infrastructure design and operations (for example, upgrade projects, architectural patterns, and so on).

So, you are probably starting to identify resources that generate costs but cannot be easily mapped or tagged because they simply depend on other resources being assigned.

Here are a couple of examples:

- **VNets**: You're not charged for the inter-VNet traffic, but you're charged for peering (between-VNet) traffic, so having two high traffic resources on different VNets may generate extra costs. However, the problem may be limited to a single application, not all the applications on the VNets.

- **Network appliances** (load balancers, application gateways, firewalls, and more): You're charged by different metrics (traffic, balancing rules, and so on), and usually, these resources are shared between different applications.

Let's analyze this simple infrastructure setup: two applications behind two different load balancers. The following diagram illustrates this setup:

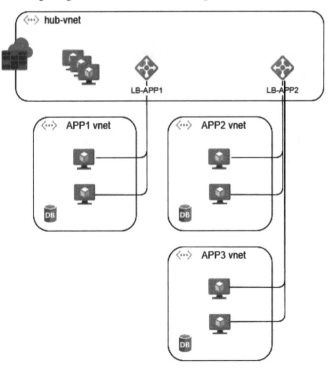

Figure 5.8 – The shared cost of network infrastructure

Here, we can see that the following:

- The application's dedicated VNets are *used* by the attached resources (VMs and **DB**).

- The load balancers (**LB-APP1** and **LB-APP2**) may be shared (or not).

- **hub-vnet** is shared between the three applications and the other services.

Fortunately, Azure allows you to get VNet-attached resources easily by API or the command line, so you can dynamically tag shared resources to split costs automatically.

Here is an example:

Resource	Criteria	Based On
Firewall	Automatically tagged	Every bsn_app tag of every object behind the firewall
hub-vnet	Automatically tagged	
LB-APP1 and LB-APP2	Automatically tagged	The bsn_app tags of the resources behind each backend pool
APP1 VNET, APP2 VNET, and APP3 VNET	Automatically tagged	bsn_app of the resources attached to the VNet

Table 5.7 – Automatic bsn_app assignation criteria for shared resources

This way, you can start identifying and splitting the cost of shared resources by application. You will also be able to understand – in case one of these costs should unexpectedly spike – if there is an application issue that is consuming more than what they are normally entitled to.

If things are getting complicated, it might be wiser to identify a different way of assigning shared resources – one, for example, that is internal to the resource, as we'll see in the next section.

Primus inter pares – finding the cost leading resource

The more you dive into analyzing shared costs, the more you start thinking that the cost of a shared resource could not be a simple mathematical average.

Let's think about the previous network example: if APP2 makes 1 Gb/s of traffic and APP3 only generates 1 Mb/s, it appears unfair to split the cost of LB-APP2 50/50. So, you need to define a way, or a driver, to realistically split the costs of shared services.

The following example shows how to split the costs of shared components based on an external driver (for example, the traffic on the load balancer to split peering costs) concerning the VNet cost components.

One good starting point is considering the network peering traffic, billed separately, as displayed in the following diagram, where peering connectivity is tied to each application VNet:

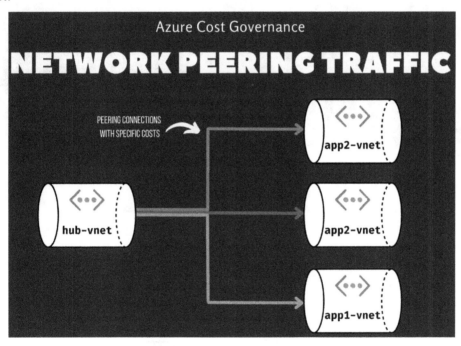

Figure 5.9 – Network peering traffic for the hub and spoke topology

By using the `.csv` invoice or the Cost Management tool, you can identify the peering traffic component of each VNet, which allows you to build the driver. As shown in the following screenshot, you can filter by **Service name** and set the **Group by** option to **Meter subcategory** to obtain the networking resources split:

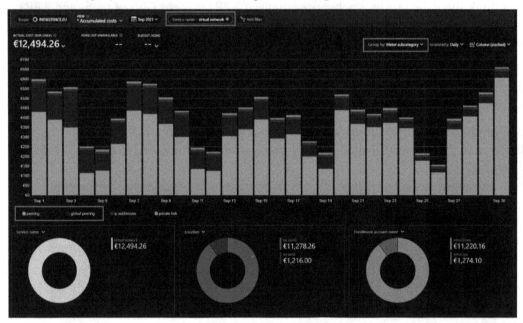

Figure 5.10 – Costs of peering connections

Let's consider the following costs for building up the driver:

Resource	Component	Cost
hub-vnet	Peering traffic	30.32€
APP1 VNET	Peering traffic	10.43€
APP2 VNET	Peering traffic	25.72€
APP3 VNET	Peering traffic	6.43€
LB-APP1		21.20€
LB-APP2		45.78€

Table 5.8 – Sample costs for network resources

LB-APP1 only serves the APP1 VNet so 100% of the load balancer costs should be assigned to APP1. LB-APP2 serves the APP2 and APP3 VNets, so we need to get the split percentage:

- The formula for obtaining the *APP2* percentage on LB-APP2 is as follows:

$$\frac{(100 * (AAP2 \text{ VNet peering traffic})}{(APP2 \text{ VNet peering traffic}) + (APP3 \text{ VNet peering traffic})}$$

- The formula for obtaining the *APP3* percentage on LB-APP2 is as follows:

$$\frac{(100 * (AAP3 \text{ VNet peering traffic})}{(APP2 \text{ VNet peering traffic}) + (APP3 \text{ VNet peering traffic})}$$

Now, we can split the LB-APP2 costs:

- APP2 costs of LB-APP2: 36.62€
- APP3 costs of LB-APP2: 9.15€

Another way to split costs is to use an *internal driver*: using a specific metric within the same resource.

Let's consider a database server VM that serves three databases and has the following cost components:

Resource	Resource type	Cost type	Shared	Monthly cost	Size	Note
db-srv-01	Virtual machine	Compute	Yes	1.053€	D32s v3	
db-srv-01_ OSdisk	Disk	Storage	Yes	20€	32 GB	OS disk
db-srv-01_MyCRM-data-01	Disk	Storage	No	40€	64 GB	MyCRM datafile disk pool member
db-srv-01_MyCRM-data-02	Disk	Storage	No	40€	64 GB	MyCRM datafile disk pool member

Resource	Resource type	Cost type	Shared	Monthly cost	Size	Note
db-srv-01_MyCRM-data-03	Disk	Storage	No	40€	64 GB	MyCRM datafile disk pool member
db-srv-01_AssetMgmt-data-01	Disk	Storage	No	20€	32 GB	AssetManagement datafile disk pool member
db-srv-01_AssetMgmt-data-02	Disk	Storage	No	20€	32 GB	AssetManagement datafile disk pool member
db-srv-01_Intranet-data-01	Disk	Storage	No	40€	64 GB	Intranet datafile disk pool member
snapshot	Snapshot	Storage	No	12€		Backup snapshot

Tablet 5.9 – Sample costs for databases and VMs

The total monthly cost of the VM (and its disks) is 1.285€, but 1.084€ is the shared cost.

Unfortunately, we cannot assign 100% of the costs to each of the related applications, but we need to split the costs based on an intelligent, adaptive driver to ensure that, once defined, it will be always correct.

One easy way to split storage-bound costs is to calculate the assignment percentage based on the disk allocation. In this example, we will have the following:

- The formula for obtaining the *MyCRM* percentage is as follows:

$$\frac{(100 * (\text{MyCRM datafile disk pool member costs}))}{(\text{All datafile disk pool member cost})}$$

- The formula for obtaining the *AssetMgmt* percentage is as follows:

$$\frac{(100 * (\text{AssetMgmt datafile disk pool member costs}))}{(\text{All datafile disk pool member cost})}$$

- The formula for obtaining the *Intranet* percentage is as follows:

$$\frac{(100 * (\text{Intranet datafile disk pool member costs}))}{(\text{All datafile disk pool member cost})}$$

The calculated percentages are as follows:

Slice name	Percentage
MyCRM	60%
AssetManagement	20%
Intranet	20%

Table 5.10 – Percentage of used resources

Using the storage-based driver and considering the dedicated and shared costs, we can assign a reasonably split cost to each application, as shown in the following table:

Component	Description	Costs
MyCRM shared	Compute, OS disk, and snapshot	650.4€
MyCRM dedicated	MyCRM disk pool members	120€
AssetMgmt shared	Compute, OS disk, and snapshot	216.8€
AssetMgmt dedicated	AssetMgmt disk pool members	40€
Intranet shared	Compute, OS disk, and snapshot	216.8€
Intranet dedicated	Intranet disk pool members	40€

Table 5.11 – Application cost allocation

> **Important Note:**
> In this example, we used a simplified setup. With a shared disk pool, you may need to switch from disk cost to database size. If you're splitting a CPU-bound cost (for example, an HPC), you may need to use the running time of the jobs, or the per-minute CPU consumption % of each job, and so on.

Cleaning up with PaaS services

In any real environment, you'll have to face the challenge of splitting costs for all PaaS, SaaS, and IaaS resources, and usually, a successful approach is to identify an intelligent driver to split the cost in terms of a percentage.

It may seem difficult to find a real driver for the PaaS services, but please keep in mind that you can always focus on the main cost-impacting variable for each service and try to extract usage values or allocation values for that metric, and then apply the ratio to the whole PaaS service spending.

In the previous example, we had disk pools for an IaaS database, and we built up a storage-based splitting algorithm. In the following example, we will learn how to split the costs of Cosmos DB.

Cosmos DB's costs are mainly related to the **Request Units** (**RUs**) that are allocated for each collection, as shown in the following screenshot (for more information about RUs, please refer to the official documentation at `https://docs.microsoft.com/en-us/azure/cosmos-db/request-units`). So, we need to build up a compute-based splitting algorithm. Let's hypothesize a Cosmos DB database with monthly spending of about **€33,518.61**, as shown in the following screenshot:

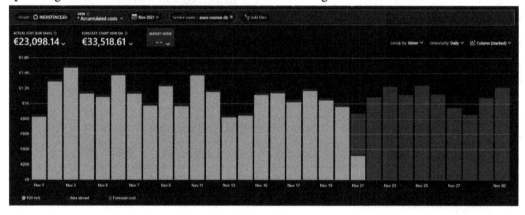

Figure 5.11 – Cosmos DB spending grouped by meter

Here, we have three collections, where each one has a manual RU assignation, as shown in the following table:

Collection name	RU
MyCRM	400
IoT	3,000
Financial	200

Table 5.12 – Percentage of used resources

So, we can calculate the impacting percentage of each collection for the entire spending process:

- The formula for obtaining the *MyCRM* collection percentage is as follows:

$$\frac{(100 * (MyCRM\ RU))}{(Sum\ of\ all\ RUs)}$$

- The formula for obtaining the *IoT* collection percentage is as follows:

$$\frac{(100 * (\text{IoT RU}))}{(\text{Sum of all RUs})}$$

- The formula for obtaining the *Financial* collection percentage is as follows:

$$\frac{(100 * (\text{Financial RU}))}{(\text{Sum of all RUs})}$$

The calculated percentages are as follows:

Collection Name	Percentage
MyCRM	11.11%
IoT	83.33%
Financial	5.55%

Table 5.13 – Percentage of used resources

Now, we can easily split the cost based on a rational driver of the total Cosmos DB instance spending:

Collection Name	Monthly Cost
MyCRM	3,723.92€
IoT	27,931.05€
Financial	1,860.28€

Table 5.14 – Final application cost allocation

We have just learned that we cannot consider a *1:1* resource-business application ratio, especially on network, compute, and shared DB or PaaS components that could represent an important slice of your monthly bill. In the next section, we will learn that unused resources also have strong security implications as they typically represent the easy and simple target for attacks.

Security implications of cleaning up

Other than simply wasting money and resources (and carbon emission, but this is something we'll address in *Chapter 8, Sustainable Applications and Architectural Patterns*), when you leave resources in a public cloud, you are putting the following at risk:

- Your virtual data center (depending on your network configuration)
- Other customers' virtual resources and data centers
- The cloud service provider's infrastructure

This goes without saying, but you need to switch off and delete unused resources, even if there is no associated cost. The best approach would be to enforce a culture of responsibility where every owner of one or more cloud resources is also responsible for deleting them.

The problem is, of course, that creation is easier than deletion for several reasons, including the fact that there are many *silent* services associated with a simple workload such as, for example, a VM (we saw this in the previous paragraphs, where creating a VM will attach several other services to it, from networking to storage, and so on). In addition, trying to control all of this without proper tagging and policy enforcement might feel like an impossible mission. Someone once mentioned that the only way of knowing what's going on in your cloud is by looking at the bill, and I think at this point of this book, we agree with this view.

Large enterprises have many reasons to spin up resources that are not typical IT business resources: they run **POCs** or design marketing campaigns or prototypes, sometimes leaving behind all the resources without proper tagging and, alas, deletion.

If this is bad at a cost management level, we must also address the elephant in the room, which is the fact that unutilized resources are the perfect targets for hackers and exploiters, so they represent the most dangerous point of entry of any organization. And if you feel that controlling your registered, accounted, and managed cloud infrastructure is an impossible task, *how about stuff you don't even know exists?* These cannot be added to your security tools and systems, so they are not scanned for vulnerabilities, patched for the latest updates, or validated against your company's policies.

So, other than simply saving money, cleaning up is vital to your organization's security posture and the recommendation of using a *zero-trust* approach also stands for orphaned resources: assume you have them in your virtual data center. Then, you can act accordingly by running audits and checking reports, and ultimately enforce a cost governance routine that will periodically take care of their deletion.

We hope that we have convinced you to get rid of all the unnecessary resources and money-spending stuff that was left in your virtual data center. However, there's one last thing you must do before considering your cost governance as *clean*, and that is handling all the unused subscriptions, as we'll see in the next section.

Unused subscriptions

In my experience, large enterprises that use the cloud for a few years typically end up with lots of subscriptions that were created for many reasons besides the official hierarchy that we suggested in *Chapter 1, Understanding Cloud Bills*: subscriptions coming from user tests, free credit subscriptions received during marketing events, prototypes, POCs, tests, and so on.

Don't get me started on how many *subscriptions* or *Microsoft Azure Enterprise subscriptions* (this is the standard default name when a subscription is created in an enterprise agreement) I have been able to find on any named enterprise customer. Keeping many unaccounted subscriptions does not necessarily bring extra costs (unless you forget resources that are still running in them), but most of these subscriptions will stay in your system with ghost resources for months and years without bringing any value to your data center.

> **Important Note:**
>
> To be able to rationalize your subscriptions, you will need to have AD administrator privileges and/or be able to access the **Enterprise Administrator (EA) Portal** (for enterprise agreements). Unfortunately, if you have subscriptions that were left from departed employees or collaborators and you cannot access them, the only way to get rid of those subscriptions will be to open a service request to Microsoft.

As explained in *Chapter 2, What Does Your Cloud Spending Look Like?*, when addressing the subscription hierarchy, there are several tools to help you handle and manage subscriptions, such as **Management groups**:

Figure 5.12 – Creating a management group to quarantine subscriptions

Once you have EA/AD admin privileges, you should be able to move the unused subscription to a specific management group (that is, *todelete*). Then, you can apply all your security and cost control policies before quarantining and deleting the unused subscriptions. This can all be done through the Azure portal: `https://docs.microsoft.com/en-us/azure/cost-management-billing/manage/cancel-azure-subscription`.

From the moment you delete a subscription, its billing is immediately stopped, and all its services are disabled (IP addresses are freed, VMs are deallocated, and so on). The storage is rendered read-only and there is still a chance to undelete the subscription if you made a mistake, but only for a few days once the action has been performed.

> **Important Note:**
> Please note that because the subscription is also a billing container, even when empty, it will take 30 to 90 days to be effectively deleted from the systems. For billing purposes, it may stay longer in your Cost Analysis tool, depending on the scope you are selecting.

With that, we have learned how to identify and remove spikes, as well as unattached, ghost, and unused objects in a virtual data center. This may not just have cost-saving results but may also make your cloud data center more secure.

Summary

In this chapter, we looked at the rest of the cost-saving techniques that you can use to both lower and keep your Azure costs down – we learned how to deal with cloud resources with the correct mindset.

Then, we learned what resources and workloads can be cleaned up and how to save money, from shared resources to networking resources consumption, down to proliferating unused or forgotten subscriptions. With the knowledge you've gained in this chapter, we hope that you are aware of all the possible hidden resources and costs of a normal operational virtual data center.

The next chapter is dedicated to reservations since they represent the quickest cost-saving technique. Even though, in complex environments, reservations may seem elaborate and difficult to deal with, we have now tackled everything we needed to properly address them: from right-sizing our resources to cleaning up unused services. Hence, we are now ready to dive into reservations with the correct approach.

Questions

1. How do you know if you have unattached disks in your environment?

2. How can you calculate the peering traffic of an application that shares a VNet?

3. How can you delete an unused subscription?

4. How can you get the costs of resources involved in a software migration project?

5. How can you split the costs of a shared resource?

Further reading

- Migration best practices for costs: `https://docs.microsoft.com/en-us/azure/cloud-adoption-framework/migrate/azure-best-practices/migrate-best-practices-costs`

- Unexpected charges: `https://docs.microsoft.com/en-us/azure/cost-management-billing/understand/analyze-unexpected-charges`

6
Planning for Cost Savings – Reservations

In the previous chapter, we learned about getting rid of unused or ghost resources and right-sizing cloud objects to their requirements. This was a prerequisite for this chapter, where we will learn how to set up a governance process to handle reservations, from understanding the billing and pricing model to successfully driving a full capacity strategy for reserved services.

We will learn about privileges, actual versus amortized views, and using a workflow for execution. We will learn what to do when usage is not optimal and understand the financial implications of reservations, as well as how to deal with repetitive changes due to the fluid nature of cloud workloads.

The following topics will be covered in this chapter:

- What happens when a resource is reserved?
- Reservation utilization
- Dealing with changes and cancellations

After reading this chapter, you will be finally ready to make your reservations' purchase, having dealt with everything you need to know and make decisions about, and be ready to handle your reservations after purchasing them for maximum cost control.

Technical requirements

For this chapter, you'll need the following:

- A computer with internet connectivity
- Access to the Azure portal
- Privileges to access Cost Management information (see `https://docs.microsoft.com/en-us/azure/cost-management-billing/costs/assign-access-acm-data`)

What happens when a resource is reserved?

Since Azure Reservations first became available a few years ago, it has quickly become a fast and easy way for customers to save money, with a time commitment of up to 3 years of usage. Depending on the type of resource, its model, and the commitment duration, the discounts can go as high as 72% (for example, on a 3-year reservation of an *M*-sized **virtual machine** (**VM**)). This can significantly cut costs and represent the first quick win of any cost management governance process.

On the other hand, customers who have embraced the cloud paradigm of fluid and ever-changing resource consumptions are always a bit wary of *committing* to a specific resource for many months. When a resource is reserved, there is nothing static about it: it is simply a billing discount that is booked in advance with the condition of spending the amount of money that the reservation is entitled to.

Please note that in non-US countries, due to the exchange rate for the related currency, you might experience small deviations from the `.csv` export in your final invoice.

> **Important Note:**
> Reservations, despite their suggestive name, are simply billing exercises and they are in no way a warranty for truly reserving a specific service capacity. The capacity setting, as we'll see later, will not guarantee that the resource will be 100% available. There is a new capacity feature that's currently in preview that will provide a reservation on capacity: `https://docs.microsoft.com/en-us/azure/virtual-machines/capacity-reservation-overview`.

Reservations can be purchased directly on the portal (or via PowerShell, the **command-line interface (CLI)**, and even through the Azure portal's **application programming interface (API)**) by paying all the amount up-front in advance, or by paying the monthly quota of each resource in the end-of-the-month billing charges. Monthly payments are only available for Azure services. The reservation process is a discount that is applied automatically at the end of the month, according to the scope, region, and **SKU** that was reserved and what has been used for that period.

The reservation discount is automatically applied to VMs and other services that match the attributes and quantity of the reservation and covers only the infrastructure cost (that is, it never covers the OS or the software part). For more information, please refer to `https://docs.microsoft.com/en-us/azure/cost-management-billing/reservations/reserved-instance-windows-software-costs`.

> **Important Note:**
> Reservations are applied on an hourly basis, except for Databricks (at the time of writing); therefore, any cost analysis to determine the right sizing and reservations should consider the hourly usage distribution of each service.

The cloud services that can be reserved started with VMs, which are probably the most popular and used, but additional services keep getting added by the Microsoft billing team regularly.

In addition to the Azure reservable services, there are discounts on third-party software plans, such as SUSE Linux, RedHat Plans, VMware Solution by CloudSimple, and RedHat OpenShift.

> **Important Note:**
> If you are using **Azure Hybrid Usage Benefit (AHUB)**, as described in *Chapter 3, Monitoring Costs*, please be aware that the reservation discount does not apply to software licenses for Windows Server and SQL Database.

As we mentioned previously, VMs were the first cloud objects available for reservations and probably the easiest to deal with, but other objects have been added as well, which we'll see in the next section.

Learning about reservations for PaaS, Storage, and data services

At the time of writing, several storage and data resources can be discounted with reservations:

- Azure Storage
- Azure Cosmos DB
- Azure Data Factory
- SQL Database vCore
- Synapse Analytics
- Azure Databricks
- Database for MySQL
- Database for PostgreSQL
- Database for MariaDB
- Azure Data Explorer
- Cache for Redis
- Azure Disk Storage (for premium SSDs > P30)

These storage resources are typically discounted based on the hourly usage and they have a *use-it-or-lose-it* policy – if, for any reason, your cloud resources have not used the discount, you will pay for the reserved object anyway.

> **Important Note:**
> As *Chapter 5, Planning for Cost Savings – Cleanup*, which is dedicated to cleaning up, mentions, sometimes, entire storage accounts are left behind as ghost resources. A good practice in such a case is to take advantage of the **last access time** feature to create a cleanup policy for storage accounts before considering reserving. You can find additional information on setting up a policy here: `https://docs.microsoft.com/en-us/azure/storage/blobs/lifecycle-management-overview`.

In addition to storage and data, some other PaaS services can be reserved, such as the **App Service** (the Azure web application PaaS service) stamp fee; more services will be surely added in the future. Before you consider reserving such services, we recommend that you use them first and understand their usage patterns, then make the break-even calculation, as described in the next section, or consider a demand shaping policy (such as the workflow for application usage patterns, as described in *Chapter 5, Planning for Cost Savings – Cleanup*). If this still makes for a good business case, then reservation is the last step to healthy cost governance for these services.

Deciding crossroads before reserving

In my experience, the first reservation purchase is like a watershed, especially for large enterprises with many departments involved in purchasing and the financial implications of it, because it opens many parallel threads with the other departments before the IT admin can click on that **Purchase** button.

As an initial requirement, to be able to purchase a reservation, you must have an **Owner** role or a **Reservation Purchaser** role on an Azure subscription, which can be Enterprise Agreement (MS-AZR-0017P or MS-AZR-0148P), pay-as-you-go (MS-AZR-0003P or MS-AZR-0023P), or Microsoft Customer Agreement. For CSP, you can use Partner Center, the Azure portal, or the Partner Center API.

> Tip:
> Please refer to `https://docs.microsoft.com/en-us/azure/cost-management-billing/reservations/prepare-buy-reservation` for more information about EA, MCA, and others, and `https://docs.microsoft.com/en-us/partner-center/azure-reservations` for CSP.

Therefore, the correct flow of steps to be taken before and after buying reservations is still the best approach, as shown in the following diagram (using VM Reservation logic flow):

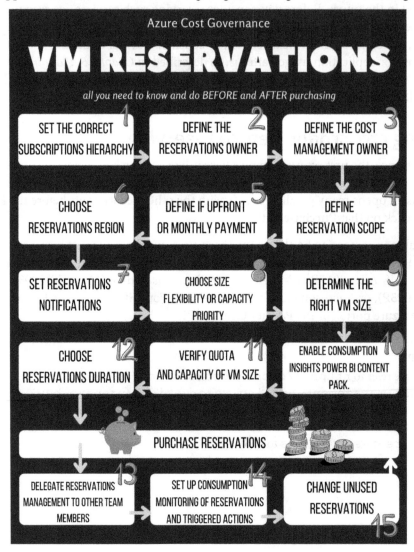

Figure 6.1 – Workflow for VM reservations purchase

As you can see, here are the steps involved in the VM reservations purchase workflow:

1. **Set the correct subscriptions hierarchy**: As we discussed in *Chapter 1, Understanding Cloud Bills*, before proceeding with any cost management technique, I recommend setting your subscription and management groups hierarchy.

2. **Define the reservation's owner**: I suggest creating an **Active Directory** (**AD**) group (for example, `Azure_reservation_admins`) to easily add or remove people in the company without having to modify every reservation.

3. **Define the cost management owner**: Owners must be **Enterprise Agreement** (**EA**) owners if the **Add reserved instance** option is disabled in the EA Portal. The Cost Management owners must own at least one subscription.

4. **Define reservation scope**: The reservations can be scoped by subscription or resource groups:

   - **Single resource group scope**: Discounts are applied only within the selected resource group.

   - **Single subscription scope**: Discounts are applied only within the selected subscriptions.

   - **Shared scope**: Discounts are applied throughout all the subscriptions on a first-come-first-serve basis.

   This can be changed after buying the reservation.

5. **Define if you will pay upfront or monthly**: Depending on the type of agreement you have, to use any monetary credit that is due to expire, choose upfront payment; monthly payments let you spread the resource quota spend month by month.

6. **Choose a reservations region**: The reservation region is mandatory due to cost constraints varying between regions.

7. **Set reservations notifications**: Only the contacts in the EA notification page will be notified, while users with the **role-based access control** (**RBAC**) configuration do not receive these notifications.

8. **Choose size flexibility or capacity priority**: Flexibility gives you the chance to use different VM sizes in the same VM size group (please refer to the official ratio list at `https://isfratio.blob.core.windows.net/isfratio/ISFRatio.csv`), while capacity will keep the VM's size fixed and prioritize it toward the data center's capacity.

> **Important Note:**
> Please note that, at the time of writing, the capacity priority is not a guarantee of capacity, but a new Priority Capacity Feature is in public preview.

9. **Determine the right VM size**: The right VM size and quantity are first suggested by the reservation page. Additional information can be gathered by Azure Advisor. Information can also be pulled from the Cost Management API and the Power BI app.

10. **Enable the Azure Consumption Insights Power BI content pack**: If you plan to manage the view of your subscriptions via the Power BI app, you need to enable the content pack for reservations.

11. **Verify the quota and capacity of VM size**: To purchase reservations, your subscription(s) must have a big enough quota for that category. You can verify this in the Subscription settings (refer to `https://docs.microsoft.com/en-us/azure/azure-resource-manager/management/azure-subscription-service-limits`) and raise a ticket to request more quotas for your subscription beforehand.

12. **Choose reservations duration**: This can be 1 year or 3 years.

    Purchase reservations: Once you have bought the reservations, there is still work to do, however, as summarized in the upcoming steps.

13. **Delegate reservations management to other team members**: Once you have made the purchase order, you can add roles for your reservation order or reservations in the **Identity Access Management (IAM)** page of each reservation. As we suggested earlier, we recommend using groups to avoid menial tasks such as manually managing users. Being the owner of a reservation is not equivalent to being a subscription owner.

> **Important Note:**
> To proceed with these changes, you need to be the owner of the entire reservation order.

14. **Set up consumption monitoring of reservations and triggered actions**: Reservations are good for cost management as they set predictable pricing with high discounts for your services. However, this only works if you then use them, ideally, close to 100% usage. Not using reservations, in addition to buying other types of services (for example, reserving a B2 VM but then using a B2ms), has the effect of paying twice for the same resource!

15. **Change unused reservations**: During your cost governance process, you must consider making changes to reservations, since you always want to reach full usage. On the other hand, as conditions in your applications and services might change, you will likely need extra capacity or decide to switch off an entire application that you reserved capacity for.

There are several reasons to make changes; the key takeaway is that if you followed our guidance, this will be much easier and more manageable in the long run. In addition, your cost controller must know that changes can be frequent, depending on your organization's resource usage.

Getting information on your objects' usage

Now that you have made a list of what you need to purchase your reservations, and assuming you have fully read, understood, and applied everything we have learned from *Chapter 4, Planning for Cost Savings – Right Sizing*, and *Chapter 5, Planning for Cost Savings – Cleanup* – that is, that all your workloads are relevant at the minimum possible size/tier and that you have cleaned up every unnecessary service – there are three ways of viewing your reservations needs within your Azure data center:

- **Reservations' recommendations**: When you select **Reservations** and click **Add**, the Azure portal will display a list of recommended reservations based on your recent usage patterns. You can drill down for each recommendation and display the savings with different options.

- **Advisor Cost Recommendations (reservations)**: If you open the **Advisor** page in the Azure portal and choose the **Costs** section, you will also find recommendations for your workloads that can be reserved, also based on your utilization pattern.

- **Reservation Coverage from the Cost Management Power BI app**: If you have installed the Power BI app, you will have an entire report dedicated to reservations, which will give you details about the breakdown of the total usage, even for resources you have already reserved.

We'll dig deeper into reservations' details in the next section, where all the possible scenarios will be presented and analyzed through examples.

Deciding whether reserving is the right choice

One of the most common questions I hear on cloud cost is: *Is the reservation worth it?* Well, fortunately, this is a math problem, and we can find out the answer without having to guess. As we learned in *Chapter 1, Understanding Cloud Bills*, all cloud resources have a metering timeframe. For example, on VMs, this is per second, but the pricing is reflected daily on public sites, and this is the information you will need to understand the *break-even* concept.

For each resource you plan to reserve, you need to find out whether the time you will keep that object switched on for makes it convenient to make a reservation. If you plan to use the resource for a time that is bigger than the break-even, then reserve it. If you are close to the break-even point, consider switching off your VM as much as you can to get below that point.

The following is the generic formula for calculating the break-even hours:

$$\frac{\text{Monthly cost of reserved resource (1 or 3 years)}}{\text{Pay} - \text{as} - \text{you} - \text{go hourly price}}$$

Formula 6.1 – Calculating the reservation break-even hours

You simply divide the *monthly cost of the resource (1 or 3 years)* by the *pay-as-you-go hourly price* and get the minimum amount of hours that the resource needs to run for its reserved price to be convenient. If you plan to keep that resource switched on for several hours that is less than the formula's result, then reserving it is more costly and not recommended.

Let's try to frame this with an example.

Example of calculating the break-even of a VM reservation

I want to run a Linux Ubuntu VM – say, an **E2as v4** instance – in my virtual data center for 8 hours every day. This means 40 hours per week and 160 hours per month.

The following screenshot shows the pricing of a Linux **E2as v4**:

OS/Software:	Category	VM series:	Region:
CentOS or Ubuntu Linux ⌄	All ⌄	Easv4-series ⌄	West Europe ⌄

Currency:	Display pricing by:
Euro (€) ⌄	Hour ⌄

Instance	vCPU(s)	RAM	Temporary storage	Pay as you go	1 year reserved	3 year reserved	Spot *	Add to estimate
E2as v4	2	16 GiB	32 GiB	€0.1282/hour	€0.0754/hour ~41% savings	€0.0507/hour ~61% savings	€0.0270/hour ~79% savings	⊕

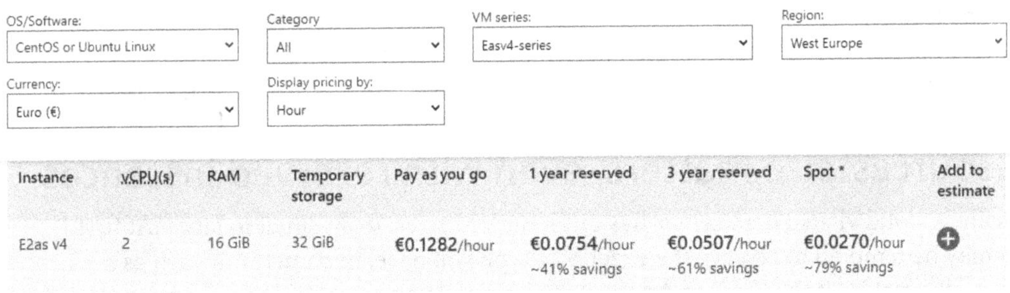

Figure 6.2 – Pricing of a Linux E2as v4, West Europe

The cost of this VM (in terms of pay-as-you-go) is 0.1282 x 160 = 20.512 Eur/month.

The cost of the same VM with a 1-year reservation is 0.0754 x 730 = 55.042 Eur/month.

The cost of the same VM with a 3-year reservation is 0.0507 x 730 = 37.011 Eur/month.

How do you calculate the break-even of a resource to understand whether it is convenient to reserve? If you divide the monthly cost of 3 years by the pay-as-you-go hourly price, you get the number of hours you will need to keep the resource on for it to be convenient.

So, in our example, 37.011/0.1282 = 288.69 hours.

If your VM is switched on for more than 288.69 hours, then you should reserve (for 3 years); otherwise, it is more convenient to pay only when consumed.

Example of reserving a PaaS object

Let's also clarify this concept regarding PaaS products, which are a bit different than VMs, and because of their peculiar pricing and tiers, different from each other. Let's imagine that you are using Azure Cosmos DB and, after a decent amount of time of stabilizing its features and usage, you decide to reserve it.

The billing you have in your monthly report is for its database operations, consumed storage, and optional dedicated gateways. You can reserve the request units per second (RU/s), starting from 5,000 RU/s, either single or multi-region write, for 1 or 3 years. Any provisioning beyond the reserved quantity will be charged at regular throughput rates.

As for the previous example, the reservation is valid until you have resources that can be discounted, while if, for any reason, you decide to switch them off, you will lose the reservation (and money) for that time.

In addition to Cosmos DB, there is a regional ratio to apply the **Database Transaction Units (DTUs)**, which can be found here: `https://docs.microsoft.com/ en-us/azure/cost-management-billing/reservations/understand- cosmosdb-reservation-charges?toc=/azure/cost-management- billing/reservations/toc.json#reservation-discount-per-region`.

Resources' size matters, even with reserved instances

Usually, a cloud virtual data center has different VM sizes: from small to large. At first, you may be tempted to reserve the exact VM type you have, in something such as a 1:1 approach, or at least this is what **Advisor** recommends.

However, you need to consider that the reservation discount is likely flat, for the reservation timeframe, inside a family. For example, all the **Fs** VMs (**F2s** such as **F32s**) have the same discount for the 1-year reservations.

As we mentioned earlier in this section, there are two main settings for VM reservations: instance flexibility and capacity priority. You can typically use instance flexibility when you have a large number of VMs that use the same VM family, where you can optimize your usage even if the VMs are not 100% switched on:

- If you are using a VM for a fraction of an hour (which is the billing chunk), and another VM of the same family for another fraction, Azure will use the reservation to cover both usages according to the time fractions, up to the billed hour and the core calculations.

- If you are switching off a VM for a flexible family that you reserved, the discount can be applied to another VM of the same family but of a different size, according to the related ratio.

Let's clarify this concept with a practical example.

Practical example of VM shared reservation

Let's imagine we have the following VMs running:

- Two E4s
- Four E8s
- Two E16s
- One E32s

Now, what happens if we buy an E32s shared reservation? There are several possible scenarios as the Azure billing backend could use your E32s reservation to cover one of the following:

- Both E16s

- The E32s

- All the E8s

- Both the E4s, one E16s, and one E8s

- A small fraction of everything!

In the following screenshot, you can see the **Utilization over time** and **Daily usage breakdown** sections of all your VMs reservations, along with the fractional quantity you can use for the flexibility option:

Figure 6.3 – Azure portal – utilization over time of my E32s reservation

In the following screenshot, you can see the **Daily usage breakdown** section:

Subscription	Resource group	Resource name	SKU used	Quantity used	Instance size flexibility ratio	Normalized quantity used
INEXISTENCE - Primary	NE-RG-MyCRM	mycrm_appsrv01	Standard_E4s_v3	17	8	2.125
INEXISTENCE - Primary	NE-RG-MyCRM	mycrm_appsrv02	Standard_E4s_v3	17	8	2.125
INEXISTENCE - Primary	NE-RG-MyCRM	mycrm_docstore01	Standard_E8s_v3	17	4	4.25
INEXISTENCE - Primary	NE-RG-MyCRM	mycrm_dbserver01	Standard_E16s_v3	17	2	8.5

Figure 6.4 – Azure portal – a fraction of my E32s reservation

As you can see, it's very hard to predict which of the objects the Azure billing backend will apply the reservation to, especially when you use shared and flexible reservations. On the other hand, the shared reservation type gives you more flexibility and ensures that your reservation will be used at the maximum.

So, if we have a high number of VMs, reserving 1:1 could overcomplicate how the reservation's coverage and optimizations are managed, especially when you start right-sizing your VMs or decommissioning an entire application.

Is there anything we can do to simplify reservation management? Fortunately, yes. We can start buying and/or exchanging (we'll learn about changes in the following sections) at the least common denominator. In the previous example, the size of **E2s** will cover all the other E family sizes.

Azure Cost Governance		
BASE VM MULTIPLIER FOR RESERVATIONS		
VM size	Multiplier	base VM size
E4s	2	E2s
E8s	4	E2s
E16s	8	E2s
E32s	16	E2s

Table 6.1 – Base VM multiplier for reservations

Let's calculate the total **E2s** of the least common denominator needed:

Azure Cost Governance				
LEAST COMMON DENOMINATOR VMS SIZE CALCULATION				
VM size	# of VMs	base VM size	Multiplier	Total # of base VMs
E4s	2	E2s	2	4
E8s	4	E2s	4	16
E16s	2	E2s	8	16
E32s	1	E2s	16	16

Table 6.2 – Least common denominator for VM size calculation

In conclusion, instead of purchasing four different reservations and complicating their management, you can safely buy 52 **E4s**, or better, 104 **E2s**, and the Azure billing backend will reassign a reservation or a fraction of it on an hourly basis.

This approach will not only simplify reservations' management but will also save you in the following scenarios:

- You are performing right sizing
- You are stopping or decommissioning existing VMs and creating VMs of the same family but of different sizes

You'll only have one reservation to manage by simply calculating the equivalent minimum denominator for each family.

> **Important Note:**
> The flexibility option has strict rules about which VM family you can use and with what ratio. We highly recommend checking the official documentation for the correct VM family calculation ratio.

Full details about reservation flexibility can be found in the official documentation: `https://docs.microsoft.com/en-us/azure/virtual-machines/reserved-vm-instance-size-flexibility`.

> **Tip:**
> Please always consider using the **Exchange** or **Refund** feature to adjust existing reservations using the least common denominator approach.

Now that you have understood the equivalent minimum denominator approach, I suggest that you calculate it periodically for all your virtual data center resources and compare it to your reservations. This way, you will quickly understand if you're reserving the correct number of equivalent resources, if you need to buy more reserved instance(s), or if you need to exchange them for new, different VMs families.

Setting reservations' privileges

Although you have understood the process of making a reservation, there are still many interesting and important things to know about it. The **reservation order** is the object that holds all the permissions, and all the objects in it will inherit the privileges from it. By default, the person who bought the reservation and the account owner of the subscription are allowed to access the reservation order (via RBAC).

In addition, people with the Billing Contributor role will be granted access by default. To allow anyone else to access the reservation order, one of the previous owners will have to grant access by either assigning an owner role to the reservation order or by adding a billing administrator user (in EA agreements) or a billing profile owner/contributor (for Microsoft Customer Agreement).

So, a reservation order can be seen as every other Azure service, where you are the owner of the newly created object and can decide the RBAC for each reservation order and reservation.

> **Important Note:**
>
> Your reservation order is the billing container of all your reservations and can contain several types of different reservations. Managing your reservation orders can be difficult once they accumulate over time. We recommend keeping things simple: making one reservation order for each Meter Category (that is, VMs) to avoid mixing different types of reservations and billing meters.

Limiting the number of reservations in a reservation order to a single family of Meter Subcategory will make it easier to control your utilization across that subcategory.

When managing reservations, you will need permissions for the Reservation Order, and this is not equivalent to any subscription permission.

> **Tip:**
>
> Since September 2021, a new preview feature allows users to share the reservation scope with an entire management group; that is, **management group scope**. In this case, the reservation discount is applied to each resource across all the subscriptions that are in the scoped management group.

Actual view versus amortized view

In *Chapter 3*, *Monitoring Costs*, we learned how to view and analyze our spending using the Azure Cost Management portal. One of the options on the Cost Analysis page was a very useful one for being able to track your spending properly with reservations: the **amortized view**.

The amortized view will take the total value of the reservation and break it down into daily/monthly quotas to account for the spending of those resources, which were discounted as per the reservation agreement. When toggling between actual and amortized views, you will see the following:

- **Actual view** will display the reservation purchase, but only the day when this action was performed.

- **Amortized view** will display costs allocated to the resources who got the associated discount.

Since understanding the **actual, amortized**, and usage charge types is very important, let's clarify this concept with an example.

A practical example for understanding actual versus amortized

Suppose that, in January, you bought a 1-year upfront VM reservation. Later, in March, you bought a 1-year monthly VM reservation.

Now it's April, and you need to know the following:

- **The costs of resource usage consumption (without any kind of reservation)**: To display this cost, we'll use **ACTUAL COST**. You need to set the filter for **Charge type** to **usage**:

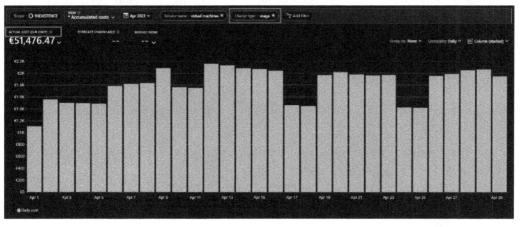

Figure 6.5 – Virtual machine usage costs

- **The costs of the resource's usage consumption and the monthly reservation, since you may consider it as a discount on the resource usage**: To find out this cost, we'll utilize **ACTUAL COST**.

This time, you don't need any **Charge type** filter. To highlight usage versus monthly reservation purchase (because you can only view the monthly one in the actual view), you can set the **Group by** option to **Charge type**, as shown in the following screenshot:

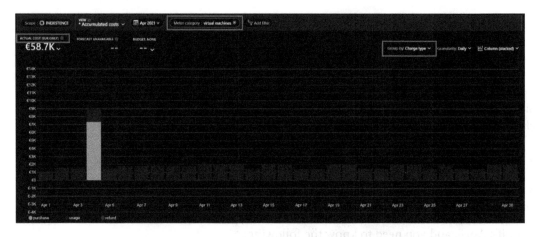

Figure 6.6 – Virtual machine usage and monthly reservation costs

- **The complete cost of every monthly reservation and the monthly accrual quota (1/12) of the upfront reservation you bought in January**: To calculate this cost, we'll use **amortized cost**. You don't need to filter by **Charge type** because you need all the charges (purchase, refund, and unused reservation).

In a very dynamic cloud environment, where you often need to exchange and refund reservations, with the preceding logic, you can always keep the costs of your reservation under control with a little math:

- Amortized costs - actual costs = the monthly competence of the upfront reserved instances.

- Amortized costs - actual costs (where **Charge type** is set to **usage**) = the monthly competence of every reserved instance (upfront and monthly).

This is very important when dealing with separate reservations' budgets.

With this initial section dedicated to reservations, we have learned about the right decisional process to correctly purchase reservations, how to calculate the break-even for when a reservation is convenient for a resource, and how to correctly display all the relevant reservations' information in the Azure portal pages, to have a complete view of what you need to buy for how long and with what specifications. The next section will be dedicated to understanding utilization to get the best out of your reservation.

Reservation utilization

As you might have figured out from the reservations' workflow, the hard work begins after purchasing them and finding out if the money has been well spent. You have committed to buying a cloud service for 1 or 3 years and in exchange, you are getting a very high discount.

But what if your internal customers forget about the reservations and start spinning up other services without correctly using those you reserved? The following steps will help us avoid this complete waste of money and resources:

1. First, you need to be able to see the cost of unused reservations. This can be done through the Azure Cost Management portal by going to the **Cost analysis** page (and even in **Power BI App**), switching to the **AMORTIZED COST** view, and setting **Meter category** to **virtual machines** and **Group by** to **Charge type**.

   As we learned in *Chapter 2*, *What Does Your Cloud Spending Look Like?*, the billing updates may take up to 48 hours to show up; hence, a good monthly report should be run after the third day of the following month.

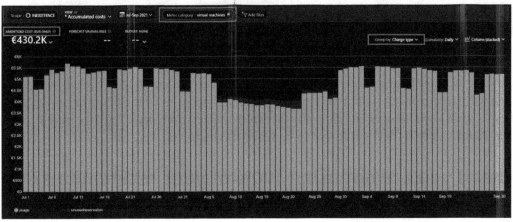

Figure 6.7 – Unused reservations in the Azure Cost Management portal

2. After that, to set the **Charge type** option to **unusedreservation**, we need to select the **Table** option so that the list can be exported for further analysis. You can also set **Granularity** and other filters as per your needs:

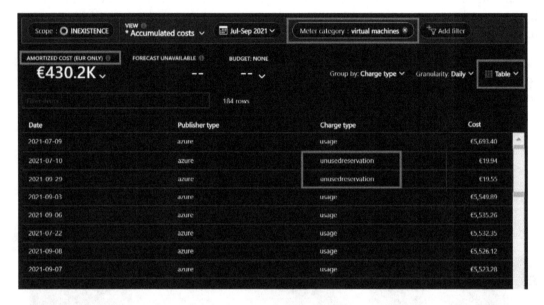

Figure 6.8 – Unused reservations in the Azure Cost Management portal in the Table view

With that, we have learned how to identify issues with reservation usage. This can be due to your services growing faster than expected, due to changes in your cloud workloads, or even, sometimes, people just choosing the wrong objects and forgetting they have a reserved option. Now, let's learn what to do when this happens.

What if you are not using 100% of your reservations?

You will need to carefully rerun the original decision workflow and understand what choice impacted your outcome:

- *Is your company using the right VM size and family?*

- *Are you using instance flexibility?*

- *Is the scope too narrow?* (for example, you might have used a resource group, but a subscription scope is wider and easier to fulfill)

Once you have found the problem, you need to make changes accordingly:

- Make sure that the used resources belong to the same family of what you reserved (for example, an Ea2 v4 is different from an E2as v4).

- Make sure you are using instance flexibility and, where possible, always buy the lowest CPU size reservation to maximize flexibility.

- Change the scope to **shared** to allow the other departments and subscriptions in your extended scope to benefit from the reservation.

- Enforce a policy of allowed SKU (we learned about this in *Chapter 2, What Does Your Cloud Spending Look Like?*).

- Lastly, consider exchanging the reservation so that it matches the usage requirements of your company.

Additional information on the official exchange and refund policy can be found here: `https://docs.microsoft.com/en-us/azure/cost-management-billing/reservations/exchange-and-refund-azure-reservations`.

Automatically renewing reservations

With all the available reservation options, you can also **automatically renew reservations**. This is particularly useful for general-purpose VMs, or database services that you know will typically only increase with time. With the organic growth of your company's cloud resources, you typically don't need to periodically assess the quantities (although we always recommend doing so) and make sure your discount is always granted. The default setting is **OFF** (disabled) for automatic renewal; you will need to manually enable it.

Right-sizing and new objects

Many customers have a tendency, once reservations are made, to forget about them because the saving was good and *there is no point in changing things that work well*. Of course, in the cloud, this is a very dangerous assumption and with reservations, I think it has a very bad impact on your saving chances. Cloud objects are always evolving, creating newer, faster, and cheaper services, and if you decide not to update those services, you are not only denying yourself some good innovation, but you are also wasting money. My recommendation is to review the right sizing (as per the process and methods explained in *Chapter 4, Planning for Cost Savings – Right Sizing*), along with reserved objects; the benefits are evident:

- You may find that your application is performing better than expected, and while downsizing a reserved service, you can make space for another application using the same resources.

- In general, for all services, especially for VMs, you may find that there are newer models that are cheaper and better performing, maybe with additional features that you did not have when you purchased the original reservation, and in this case, an exchange can be a lifesaver.

- You might be looking for an extra budget for a new project and find out you already have reservations that can cover that project with little effort regarding right-sizing.

The key takeaway is that you should never consider a reservation as static – instead, you should periodically check, update, and improve your services, as well as all the related reservations, to get the maximum efficiency out of them.

Financial implications of reservations

While buying a reservation, it's technically easy if you follow the previous steps. Please consider that you're *investing* money for a relatively long timespan over 2, 3, or 4 years within your company's fiscal range, depending on when you buy the reservation. This can also be extended when changing reservations (please remember that when you exchange a reservation, the new reservation on the new items will start from the day you made the change). In the overwhelming majority of the large-scale cloud users, you will end up reserving something.

This process should be carefully explained and shared with your cost controller because of the following reasons:

- Some companies may consider reservations as **Capital Expenditures** (**CAPEX**) and not **Operation Expenditures** (**OPEX**). So, your budget should reflect this point.

- Some companies need you to rediscount the monthly competence over the years, and you need to track down exchanges and monetary commitment.

Let's look at an example to clarify this concept.

A practical example of financial reservations

On March 13, 2021, you bought a VM reservation (for 3 years) for 10 M64s VMs, and you chose the upfront payment reservation type (let X be your spending).

In the final March 2021 invoice, you will find the total cost of the reservation, but your coverage is over a total of 4 years (2021, 2022, 2023, and 2024):

Azure Cost Governance
RESERVING 10 VMS FOR 3 YEARS

	13 Mar 2021	13 Mar 2022	14 Mar 2022	13 Mar 2023	14 Mar 2023	13 Mar 2024
10 M64	YEAR 1					
10 M64			YEAR 2			
10 M64					YEAR 3	

Table 6.3 – Reserving 10 VMs for 3 years

In this case, you must share with your cost controller or financial team that you're spending on March 2021 for resources that will cover several years. Also, since the cloud is on a subscription/pay-as-you-go model, it's an OPEX expense and needs to be rediscounted in the months.

Now, since a cloud environment can't be stable in services and tiers for 4 years, let's imagine that in 2022, you have a change in terms of the capacity requirements because of an additional workload on your application, and you calculate that you will need to exchange three VM reservations from **M64s**, 3-year, to four **E64as**, 3-year.

Now, what happens to our reservations (and investments)? Let's summarize these operations in a simple table:

	13 Mar 21	13 Mar 22	14 Mar22	13 Mar 23	14 Mar 23	13 Mar 24	14 Mar 24	13 Mar 25
Azure Cost Governance — EXCHANGING/EXTENDING RESERVATIONS								
10 M64	YEAR 1							
7 M64s + 4 E64as			YEAR 2					
7 M64s + 4 E64as					YEAR 3			
4 E64as							YEAR 4	

Table 6.4 – Example of exchanging reservations and extending the timeline

Here are some additional considerations on this purchase:

- You exchanged your reservations, ending three M64s reservations that originally ended in 2024.
- You recycled some of the initial investment for specific needs to match the current technical needs, and this is good and right.
- The new 3-year reservation for the new VM size starts in 2022 and ends in 2025.

Therefore, you'll end up with two consequences:

- In 2022, you'll recycle part of the original upfront investments to a new destination.
- In 2022, you'll need to add some money because 2 years of the three M64s reservations aren't sufficient to buy four E64as VM reservations for 3 years.

You need to share this exchange with your cost controller team because it will change the rediscount of the initial upfront value and include the difference for the new E64as reservations in 2022.

The information you are using for this analysis can be found (unaggregated) in the **Cost analysis** tab of the Azure Cost Management portal. Then, set the following parameters and fields:

- Set the **Add filter** option to **virtual machines.**
- Set the **Group by** option to **Meter Subcategory** (to obtain the VM sizes).
- Set the cost to **Amortized cost** (so you can see the discount quota for the reservations).

Then, create a second view in a new **Cost analysis** page and set these fields as stated:

- Set the **Add filter** option to **virtual machines.**

- Add a filter for **Reservation order.**

You will have to extract the data (exported via an Excel or a .csv file) and copy it to our example table, as depicted previously.

> **Important Note:**
>
> Most customers will choose the monthly payment option to make the OPEX calculation easier. Some customers have internal policies that allow the 3 years reservation to become part of the CAPEX budget (since it spans multiple years). If you plan to exchange a reservation from 3 years to 1 year, the exchanged amount may switch to OPEX, and the cost controller team should be informed.

In this section, we learned what reservations are, what decisions you will need to take, and what needs to be done to obtain the best possible outcome, which is 100% utilization of the money you invested to get the reservations discount. The last part of dealing with reservations is issuing cancellations and refunds, which we'll address in the next section.

Dealing with changes and cancellations

There are several things you might want or have to change in a reservation. The possible changeable items, according to the official documentation (https://docs.microsoft.com/en-us/azure/cost-management-billing/reservations/manage-reserved-vm-instance), are as follows:

- Scope (shared or single)
- Optimize settings (flexibility or capacity)
- Split a single reservation into two

After you click on **Purchase**, two objects are created: a Reservation Order and one Reservation. Any subsequent action that's performed on the reservation will have the effect of adding new reservation items under that original order.

Once the reservations have been purchased, you can still make changes and modifications with some specific limits that are enforced by Microsoft (namely, a refund of up to 50,000 USD on a 12-month rolling period for changes and cancellations), which makes using reservations a bit easier.

When making changes, the new amount must be greater than the refund amount: a smaller amount will prompt an error.

> **Important Note:**
> When swapping a VM reservation from a size that does not support premium disks to one that does, such as Eav4 to Easv4, this will not be equal to a refund and repurchasing it and will not reset the terms of the reserved instance.

When performing a reservation exchange, the existing reservation is canceled and refunded (for the remaining quota of the original amount), and a new purchase is added. For EA, the refund is added as credit and subsequently spent with the new items.

> **Important Note:**
> With reservations, you can only exchange when the objects are similar: VMs with VMs, even from one region to another, a SQL database with SQL managed instances, from 1 year to 3 years, or vice versa, and so on. You cannot swap objects that are of different types.

At the time of writing, some objects cannot be refunded: please check the official Microsoft documentation to check whether your purchase is eligible for a refund.

Exchange and refund policies

For all details on the cancel, exchange, and refund policies, please check out the official documentation at `https://docs.microsoft.com/en-us/azure/cost-management-billing/reservations/exchange-and-refund-azure-reservations#cancel-exchange-and-refund-policies`.

A few topics are worth noting regarding this book:

- Flexibility is only valid among VM families indicated in a specific list (as quoted in the previous sections).
- Every time you exchange a reservation, the timer on the new choice of object is reset.
- There is no penalty for exchanges, but the new reservation amount must be greater than the exchanged one.
- For refunds, there may be a 12% termination fee that Microsoft is not enforcing at the time of writing.
- Only the owner of the reservation order can issue a refund.

Considering everything we have learned in the previous sections, you can now follow up with reservations exchanges at the pace of your virtual data center and resources: my recommendation is to set up automation wherever possible to avoid human mistakes, but also to check regularly that the reservations' usage is worth your spending.

> **Important Note:**
> Please note that for CAPEX chargebacks, if your cost controller chooses to treat reservations as CAPEX with upfront payment, you may work with the monthly accrual of the used reservation. Also, when dealing with exchanges, treat the accrual only for the exchanged part related to the final chargeback entity, while reinvesting the difference on other departments, along with their chargeback.

With this last concept, we have now learned about every aspect of the reservations purchase process, which can help you save money and guide your company toward fully automated cost governance.

Summary

In this chapter, we learned how to understand the reservation process and calculate whether it is worth reserving, what decisions are required before making a reservation order, and how to successfully track your reservation usage.

We also learned how to make proper changes and cancelations when the usage is too far from the optimum (which should ideally be around 99.99%). This will provide the best possible impact on your reservation and make this operation a clear and structured process that you can share with your finance controller. This is critical, especially in companies where you need to chargeback cloud costs and have a dedicated financial controller dealing with cloud costs.

Once you are done with right-sizing, cleaning up, and reservations, the only other way of saving money is to optimize your applications, for them to consume fewer resources and spend less. This is the main topic of the next chapter: having learned how to optimize your infrastructure's costs, we will focus on optimizing your application so that it consumes fewer resources, and you will be able to go back to right-sizing, cleaning up, and reserving in your virtuous, continuous cost optimization process.

Questions

1. How can you find out whether a reservation is worth it?

2. What is the amortized view?

3. Why does your cost controller need to know about reservations?

Further reading

- Determining what reservation to purchase: `https://docs.microsoft.com/en-us/azure/cost-management-billing/reservations/determine-reservation-purchase?toc=/azure/cost-management-billing/reservations/toc.json`

- Permissions to view and manage Azure reservations: `https://docs.microsoft.com/en-us/azure/cost-management-billing/reservations/view-reservations?toc=/azure/cost-management-billing/reservations/toc.json`

- Viewing reservation utilization after purchase: `https://docs.microsoft.com/en-us/azure/cost-management-billing/reservations/reservation-utilization?toc=/azure/cost-management-billing/reservations/toc.json`

- Reservation discount: `https://docs.microsoft.com/en-us/azure/cost-management-billing/manage/understand-vm-reservation-charges`

- Troubleshooting reservations: `https://docs.microsoft.com/en-us/azure/cost-management-billing/reservations/find-reservation-purchaser-from-logs?toc=/azure/cost-management-billing/reservations/toc.json`

- Dealing with refunds and exchanges: `https://docs.microsoft.com/en-us/azure/cost-management-billing/reservations/exchange-and-refund-azure-reservations#how-to-exchange-or-refund-an-existing-reservation`

Section 3:
Cost- and Carbon-Aware Cloud Architectures

In this section, you will understand how database and application architectural patterns have a direct impact on cloud spending and what the most used strategies are to optimize performance and costs. The discussion will shift from savings on infrastructure to examine how a well-designed and performant application can bring down cloud costs and why any investment in performance optimization strategies, among them database tuning, is easily and quickly repaid by the monthly savings gained.

This section will also focus on sustainable software engineering and how an application designed for performance and cost management can have a lower carbon footprint. Upon completion of this section, you will have a good understanding of the most recent performant, cloud-native, and sustainable architectural patterns for a cost-managed and carbon-aware application.

This section comprises the following chapters:

- *Chapter 7, Application Performance and Cloud Cost*
- *Chapter 8, Sustainable Applications and Architectural Patterns*

7

Application Performance and Cloud Cost

After dealing with the most common cost-saving techniques in the infrastructural part of this book, in this chapter, we will discuss how a well-designed and high-performing modern application can play an important role in optimizing your cloud spending.

We will also look at why any investment in performance optimization strategies, such as **database tuning** and refactoring, can easily and quickly be repaid by the monthly savings, and how a strong motivator for modernization can be financial, on top of the usual technical reasons such as architecture, software updates, and so on. An application that is not fit for the cloud may bring unnecessary costs, and we will learn what the main optimization focus items are by looking at practical examples about how this approach can bring consistent down-sizing to the starting infrastructure.

In this chapter, we will cover the following topics:

- Optimizing your database for costs
- Application performance optimization
- Practical examples

Upon completing this chapter, you will have a good understanding of a database tuning strategy, as well as of what **technical debt** is and how it can be optimized to increase performance and decrease infrastructural costs.

Technical requirements

For this chapter, you'll need the following:

- A computer with internet connectivity

- Access to the Azure portal

- Privileges to access Cost Management information (see `https://docs.microsoft.com/en-us/azure/cost-management-billing/costs/assign-access-acm-data`)

Optimizing your database for costs

We often see that customers migrate to cloud applications that rely on old databases. Sometimes, even cloud-native applications are developed using old patterns for data handling, mostly because companies have a history that needs to be retained and cannot be wiped out by a new database or application.

But an old and stratified database has its drawbacks: queries are slow and resource-intensive, so typically, the reaction is to add more resources and scale vertically, which is not what this section is about. You need to consider optimizing your database so that your application is leaner and faster, but mostly so that you will save money by downgrading the infrastructure. Database performance is commonly correlated to the following infrastructural parameters:

- Read/write **input/output operations per second** (**IOPS**)

- Disk throughput

- Disk latency

- CPU and RAM

- Queue depth

This is true not only when you are dealing with IaaS compute resources directly, but also with PaaS database services, where optimization can help you change tier and model.

A combination of the preceding parameters will result in a specific VM size or PaaS service tier that may or may not (depending on the specific database requirements) overprovision one of the variables to meet the requirements for the other. Now, let's analyze the IaaS approach, which is slightly more complex, and then apply the same logic to PaaS database services. The database-hosting VMs would typically support Premium SSD and/or Ultra disks, and each size will have its limits in terms of scaling, IOPS, bandwidth, and maximum attached disks. So, you will need to choose the VM based on disk number and performance. Even if you attach powerful disks or the newly added burstable disks (`https://docs.microsoft.com/en-us/azure/virtual-machines/disk-bursting`), typically, the VM will drive the maximum performance for that workload.

For example, using a **Standard_D8s_v3** VM (which is a general-purpose VM), you get 128 MB/s of disk throughput and a *burst* up to 400 MB/s for a short time (30 minutes). If you attach three P40s (these can go up to 250 MB/s each), you are still capped at 400 MB/s for a short time, and at 128 MB/s on average.

> Tip:
> You can find VM performance information in the public documentation, by type and family: `https://docs.microsoft.com/en-us/azure/virtual-machines/sizes`.

The previously mentioned limits are only for the disks and not for network performance. Each VM has a maximum performance on the network throughput as well, and we will learn how to choose between network and storage in the upcoming sections.

In addition to the VM size, Premium Storage disks have specific increases in performance according to the storage size, which, in many cases, forces the customer to add more disks (hence, overprovisioned storage) to reach the required IOPS.

An inefficient database can result in very high CPU usage of some databases: in some cases, database VMs can be at 100% CPU usage for peak hours (mainly due to **iowait** time on disk queue – the time that the processor/processors are waiting), and this is usually due to the following concurrent reasons:

- Unoptimized database queries
- Higher disk latency
- Throttling at the disk level, which (in applications that use a plain retry pattern) will introduce forced latency on the disk

In some cases, the VM's size might be increased only to accommodate a higher number of disks, solely for performance reasons. This means that optimizing the database, improving the disk latency and throughput, and reducing the number of disks can have a side effect of lowering CPU requirements, hence lowering the VM size.

For PaaS database services, the database tuning process will depend on the type of application and might need some refactoring to tweak the performance to optimal values. If an application is *chatty*, meaning that it is making extensive use of data access operations across the network, this will not only increase costs within the database tier, but in the case of network peering, it might also increase network costs.

Alternatively, if the database has compute requirements that exceed the highest tier (in Azure SQL, for example, the Premium size), optimizing with sharding or partitioning can allow you to scale it down to a single resource. In some cases, even adding more compute resources will not fix performance issues: poorly written queries are seen everywhere, and getting rid of those is a win-win. Last but not least, applications that use deadlocking patterns might require higher compute tiers for no good reason, and a simple caching service might solve the problem and make it faster and cheaper.

You need to carefully plan a database optimization task that's embedded in your cost governance model, and then transform it into a continuous process to have it under control, not just cost-wise but also performance-wise. In the following sections, we'll learn a few tips about tuning our database with the final goal of reducing its infrastructure's footprint and costs.

Disk tuning

A common way of tuning your disks' performance is via **disk striping**, which is an OS feature where disks are divided into blocks and data is spread across them in a **Redundant Array of Independent Disks (RAID)**. In on-premises data centers, RAID has always been used for redundancy because if a physical disk fails, this won't compromise the storage and the disk can be replaced quickly. But on Azure, we already have redundancy using managed disks at the local level, as three copies of the data are written in the data center. So, striping has been purely left to performance optimization at the OS level.

I hope the simplified scheme shown in the following diagram will clarify how striping will affect data distribution on four P30 disks, assuming there's statistically homogeneous data distribution and access:

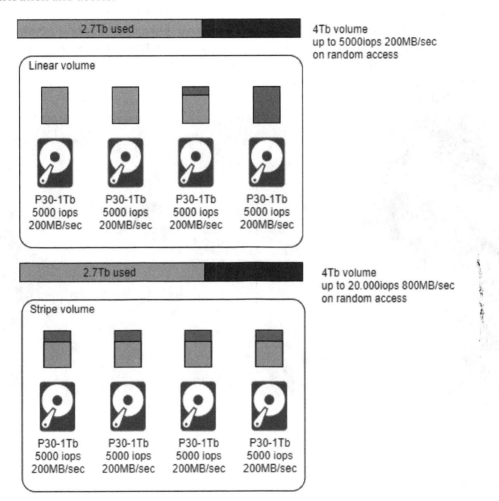

Figure 7.1 – Data distribution with disk striping

In a linear volume, data is continuously written from one disk to another. Assuming we have an initial empty volume, the first write operation will start at the first filesystem block of the first disk and will linearly continue until the end of the first disk. Then, the operation will move on to the first block of the next disk, and so on. In a striped volume, data is divided into chunks, and each chunk is stored on a different disk to take advantage of parallel access, and in turn, multiplying the total available IOPS and throughput. Let's try to figure out a way to determine the correct size of the disks to build a striped volume. The following graph shows the IOPS that was reached by striped volumes built with each model of E disks:

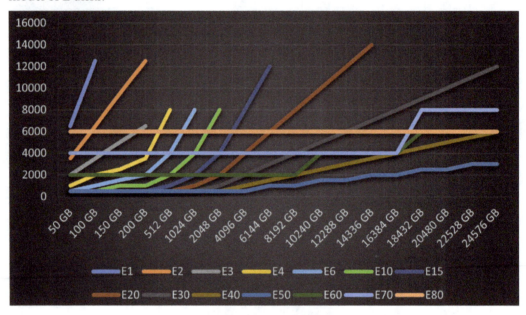

Figure 7.2 – IOPS reached by striped volumes

As you can see, an array built on smaller disks can provide more IOPS, but the VM limit on the number of attached disks may negatively affect the total size of the array since you might have to oversize the VM to accommodate more disks.

For example, an average VM does not allow you to attach more than 31 data disks. So, with E1 disks, you can reach only 124 GB, but with E2 disks, you can double the size while maintaining linearity between VM size and IOPS. Larger disks surely provide more space, but IOPS does not increase with the same ratio as the space. So, you should adopt striped volumes with smaller disks to maximize IOPS and throughput: E15 disks provide only 500 IOPS every 256 GB.

In conclusion, you need to carefully plan and then monitor the disk's usage from the Azure portal to discover if you can use a better configuration with cheaper disks or increase performance by choosing the correct striping configuration.

> **Important Note:**
>
> The closer you are to the limit of attachable disks, the more difficult it will be to reconfigure your striped volumes. So, please consider data growth and contingency-free space in advance without over-allocating storage, which is against the purpose of this whole book.

Exploring the space between the database and the application

Usually, modern applications rarely execute manual queries but are based on **object-relational mappings** (**ORMs**): a software layer that maps database query results (also known as ResultSets) to objects in the chosen programming language. The ORM libraries allow developers to write queries using the object-oriented paradigm of their preferred programming language, rather than using SQL.

Today, ORMs can be very complex with different layers of abstraction, caching, query parameterization, substitution, padding, pagination, and other features. I suggest that you invest some time checking out your ORM configuration to spot possible optimizations.

Please remember that every ORM is a complex exercise, and, like every piece of software, may have bugs that you need to fix or that will force you to upgrade the ORM and its dependencies to have the query optimization process work.

Database optimization plan

Optimizing a database won't be easy, as you might have already guessed, and you will need to work with your application owners to decide the best moment to drive changes in your databases. As we learned regarding other topics in this book, we always recommend a structured approach that can be summarized with the following workflow for database optimization. The good news is that not only will you be able to save money, but your application's performance will also improve, and its database will be healthier and more controlled.

We already mentioned the structured approach to changing less important environments (development, testing, and so on) to carefully verify and anticipate any possible side effects of the optimization process. Since, in this step, we're also modifying the application and data structures, it's exceedingly important to anticipate any eventual side effects. The following diagram summarizes the process of database optimization for an application:

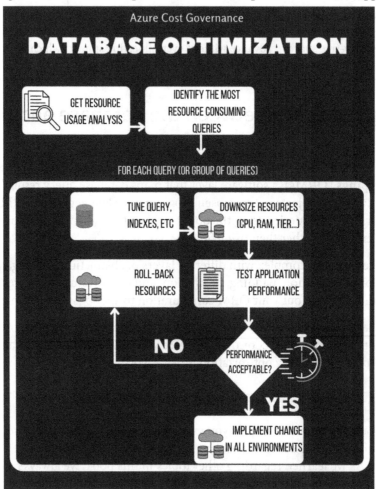

Figure 7.3 – Workflow for database optimization

Once you are satisfied with your database tuning and have right-sized your IaaS and PaaS database resources to a correct level of usage and performance, the next step of our cost governance journey is looking at the application's performance, infrastructure, and cost.

Application performance optimization

Having fully optimized your databases against your infrastructure is, unfortunately, sometimes not enough. If your application is running in Azure, it probably came from one of the most common migration patterns:

- **Lift and shift**: You took the application and moved it to cloud resources without changing anything. This is typically the most expensive pattern infrastructure-wise as matching on-premises resources within the cloud is a time-consuming and often useless activity in terms of modernization, since you will often just move your legacy app somewhere else.

- **Refactor**: You made small changes in the code and probably took advantage of a few PaaS products and now have a mix of old and new cloud objects, as well as a roadmap to something more modern.

- **Rearchitect**: Chances are you cannibalized the original application toward a microservices solution, which is much better than the preceding two patterns, but even if this dramatically reduces the infrastructure costs, sometimes, it can turn out to be expensive application-wise in the long run.

- **Rebuild**: You recreated your application using cloud-native architectures and technologies and now have a brand new, shiny workload that needs very little tuning (but still must keep moving, being in the cloud).

> **Important Note:**
>
> When migrating legacy applications, either from on-premises data centers or from IaaS, Azure offers a service named **Data Migration Assistant (DMA)**, which can play a key role in modernizing your application. DMA will help you assess, discover issues, target new features of the Azure cloud data services, migrate schemas and databases, and offer guidance on how to modernize your application. You can find additional information here: `https://docs.microsoft.com/en-us/sql/dma/dma-overview`.

The following diagram shows the various application cloud migration patterns:

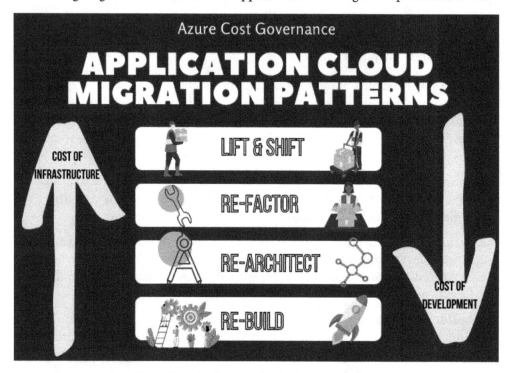

Figure 7.4 – Application cloud migration patterns

In all cases, your application needs reworking, from moving out of the VM and compute IaaS resources up to the newest cloud-native pattern. This is because a continuous improvement governance process is necessary to keep performance and costs in perfect balance. Let's analyze the concept of technical debt, which is useful for fulfilling a healthy cloud cost governance process.

Understanding technical debt

When applications are outdated and their architecture remains unchanged for many years, you will end up with code that will need heavy refactoring and old architectures that are perhaps unsupported in cloud environments. On the other hand, cloud-native architectures start with the coolest, newest patterns and cloud platform resources, but many customers forget that a cloud provider, on average, will add a thousand new features per year, and as a result, they might become obsolete very quickly. Another reason for such change is the organic growth of an application that might exceed its original conditions and will render the initial assumptions invalid.

> **Tip:**
> The technical debt is the work that must be done to refactor and improve your application, be it legacy or cloud-native, by roughly estimating the time needed to fix things that are not right and change architectural patterns that have evolved. It is measured by the number of men/days of effort to fix things.

Sometimes, we make decisions on the life cycle of an application based on several variables, such as time to market, cost of deployment, and so on. We should always consider that if we don't fix what is wrong, this will not simply go away – it will grow with time and become a much bigger or broader issue later.

Finding a KPI for your technical debt

Since it's very important to have a measure of the technical debt for an application (it's also directly connected with the application's vulnerability and business impact's risks), we need to find a way to assign a *technical debt score* to each application.

I can suggest at least a couple of ways to calculate an index to be used as your technical debt KPI:

- The number of months to rework your application to a modern state
- The number of versions for your application to reach an updated state

These are the metrics that separate your target and the updated and modern application from your older, less updated, or maintained library, package, and dependency.

Every representation has a drawback:

- You may have a recent legacy application that's been developed with old packages that completely changed the API and now require complete refactoring.
- You may simply have the last version of a discontinued product.

So, this KPI will have to be adapted to your application catalog, and you may end up using them both, depending on the type of application. This will still be a valid exercise to keep track of your progress.

For example, an application that uses our proposed KPI notation can have a technical debt of **6m, 2.4v**, meaning that it takes **6 months and 2 major, 4 minor versions** to have the most updated and recent version for that app. Defining the KPI in terms of either versions or effort time depends on the velocity of your programmers and teams.

Of course, both the effort duration and/or versions must be weighed against resource costs and other variables in your organization, such as the project's duration or the time to market, to be useful as a benchmark of your technical debt over time. We recommend that you do a thorough assessment and analysis of how your company develops (or even just buys) applications to have a clear idea and setup for your technical debt scenarios.

In your cost governance process, this is a key element of the decision, since you can match the technical debt KPI to the current infrastructural cost. Then, you can see if you have a savings budget that can help you reduce the technical debt by optimizing the application costs.

Reducing your technical debt

By now, it should be clear that technical debt is something that will stay with our applications throughout their life cycles, but we would like to keep it to a minimum, where it makes sense. This will be explored in more detail in the next chapter, where reducing technical debt in a virtuous continuous improvement process will have additional benefits:

- Faster, smoother applications
- Increased speed when refactoring
- Decreased platform/infrastructure costs
- Lower carbon footprint

As for many other topics, we thought a summary of the considerations, as illustrated in the following diagram, might help you tackle the harsh reality of accumulating technical debt and trying to get rid of it, one application at a time:

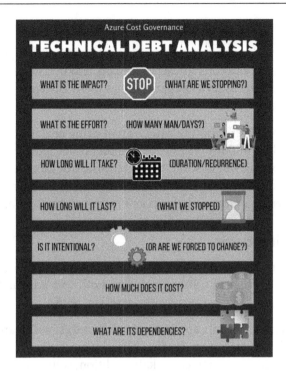

Figure 7.5 – Technical debt analysis

Other factors may influence the prioritization of an application's technical debt. For example, it might be an old software that we are decommissioning in a few months, the dependency with other applications render it much bigger and complex than it seems, or the probability of its impact is very remote. Dependencies should always be weighed in when assessing feasibility as they may hide an entire refactoring project that you did not anticipate.

Ways to optimize your technical debt and modernize your applications

So far, you have learned what technical debt is and what decisions you need to make to reduce it or at least keep it under control. Now, let's see what actions will improve and modernize your application and, as a result, your infrastructure or platform spending for it.

Microservices

A monolithic application is somehow today considered legacy, although I can think of many examples of legacy applications still serving many end users at scale through cloud services! But moving to a **microservices** approach has many benefits, from scalability to performance to versioning.

One key element for our spending optimization is that being able to scale microservices makes it a lot cheaper when microservices have different usage patterns and can scale on their own, rather than forcing you to scale the entire infrastructure. So, refactoring your application toward microservices will make it a lot easier to deal with automatic scaling and will increase your savings from a platform and infrastructural point of view.

Event-driven architectures

Serverless services (Azure Functions, Azure Logic Apps, Azure Container Instances, and so on) are among the most efficient software patterns, both cost-wise and carbon-wise: **event-driven architectures** work perfectly with serverless ones.

We are starting to see serverless-only application architectures, which are the perfect way of creating event-driven microservices that have the most efficient, cheap, and optimized software that an application can ask. In addition, solutions that use **Distributed Application Runtime (DAPR)** (`https://dapr.io/`) are starting to be embedded into PaaS services to make it easier to develop cloud-native architectures.

Output optimization and compression

In modern applications, the integration with other systems is usually implemented via `http/https` requests, and data is exchanged in a common format such as `.xml` or `.json`.

These points may sound trivial to you, but please consider that every wasted byte for every transaction is avoidable extra spending.

The first point is that humans need the output to be *pretty-printed*, with indentation and so on, but applications and servers don't. So, in your application output, you can safely delete all the indentation and formatting, saving a lot of characters (and, in the end, bytes). This will reduce the traffic and reduce the egress and VNet peering (if any) costs.

Let's consider the following `.json` notation (*we deliberately added returns, newlines, and dots to highlight the blank spaces*):

```
[\r\n
..{\r\n
```

```
....."key1":.1,\r\n
....."key2":.4,\r\n
....."document_code":.."A443"\r\n
..},\r\n
..{\r\n
....."key1":.8,\r\n
....."key2":.98,\r\n
....."document_code":.."A887"\r\n
..}\r\n
]
```

Total message length: 170 characters.

Now, the same message output can be reduced to 88 characters, with a saving of 82 characters, which is almost *50%* of the payload. Here is the same JSON reworked:

```
[{"key1":1,"key2":4,"document_code":"A443"},{"key1":8,"key2
":98,"document_code":"A887"}]
```

A second possible optimization is rethinking the key names, reducing it to 62 characters with a saving of about *63%* of the payload:

```
[{"k1":1,"k2":4,"docc":"A443"},{"k1":8,"k2":98,"docc":"A887"}]
```

In a 500 KB transaction payload, we will save about 250 KB by optimizing the output format and 315 KB with keys/content optimization. With one million transactions per day, we will save tens of GBs of traffic, as well as the related cost.

> **Important Note:**
> Please consider that human-readable CSS or comments in HTML are avoidable extra costs too! It's very important to minimize the content of scripts, tags, CSS, and every file that would generate billable traffic for you.

As you may have spotted in the previous example, in real life, we may have a lot of *talk* in the application-to-application communications that are repeated (keys, node names, attribute names, or even value substrings). In this case, we should consider enabling compression inside our application or web server (for example, using the well-known **Gzip**, a tool that is well-supported and can often be enabled without even restarting the infrastructure).

You may find it very difficult to optimize the output and activate compression on all the layers, or maybe you can't modify anything because of software support contracts or other constraints. In such a case, you must try to concentrate on the most important layer where you can save on the traffic: in a hub and spoke network topology, for example, as shown in the following diagram, there is a hub and its egress traffic.

> **Important Note:**
> This example will work for both IaaS and PaaS apps, at least with services where you don't have a native compression mechanism.

In the infrastructure shown in the following diagram, we are optimizing the compression and payload to reach the minimum message length and save on the messages:

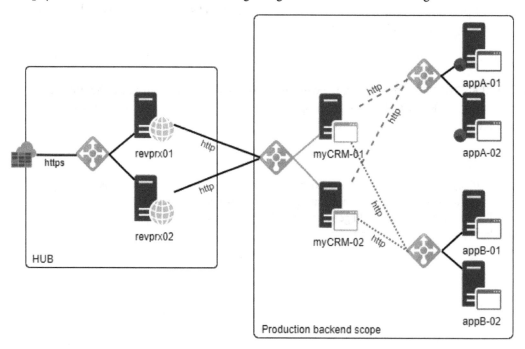

Figure 7.6 – Sample infrastructure with payload optimization and compression

Here, we have activated compression and optimized the output on the blue **appA** application servers, but you can't modify anything on the red **appB** servers.

If you analyze the traffic on **revprx01** and **revprx02** (they can be servers or appliances), you can enable compression directly in the last step before returning the data to the client.

> **Important Note:**
>
> Compression will generally increase the CPU usage by a few percentage points: you need to enable it gradually and continuously monitor the performance of the servers/appliances to avoid service slowness, performance impacts, or disruption.

Monitoring tools

When you're dealing with cloud-native applications using PaaS platform products, **monitoring tools** are essential for continuously improving and optimizing the software against fast-paced changes in terms of cloud features and adapting to organic growth and new context.

Measuring what and how the application is doing is vital to understanding what happens when an issue arises, but also to keeping its cost under control and avoiding the crazy autoscaling that I have sometimes seen in some environments that were out of control. Application performance management tools (such as Azure Application Insights) are key to having a full understanding of how you can optimize and refactor your application without impacting or, even better, improving its performance.

Application optimization and continuous improvement

A healthy application governance process should be able to continuously assess and improve the following:

- **Performance**: Ensure that the platform and infrastructure are optimized and have the right amount and quality of resources.
- **Compliance**: Make sure that your company's constraints are met.
- **Costs**: Keep your costs under control, reducing both OPEX and CAPEX.

In a traditional IT approach, efficiency and cost have historically been neglected in favor of compliance and performance, especially for mission-critical applications, at the top.

The application performance tools that I have seen usually target the highest SLA and user experience possible. But every solution is composed of multiple layers/stacks and all the layers below the application layer should be supporting the top layer.

After, you might ask reading this section: *But how much can we save with database tuning and app optimization?* Well, the only possible answer is that it depends on the situation at hand.

There are many variables at play: the type of application, infrastructure, original sizing, evolution over time, and in general, how the application was conceived to grow and evolve. These variables will lead you to either optimize and have little saving (it's still worth it, for the sake of technical debt) or find out that you had some hidden correlated spendings impacted by your application's architecture and end up with loads of saving with just the first round of optimization. Our main recommendation, as usual, is to start the process as soon as possible.

The antipattern for the brave

Depending on your application, the software developer could have implemented ORM and various kinds of design patterns, such as the following:

- Singleton
- Proxy
- Lazy initialization
- Factory
- Decorator
- Façade
- **Model-View-Control (MVC)**

Every pattern brings standardization and code clarity but may bind you to some constraints that can have adverse effects on the infrastructure. For example, in the case of one million query results, some patterns may let you load *only what is needed*, while others may require a full load and object creation before you can proceed. When you're dealing with application performance optimization, usually, there is no *best way* in absolute terms, but several sub-optimum scenarios can *carry a bit of risk*. However, they can have some positive side effects on infrastructure, performance, and user experience.

Exploiting user inactive time

Just as an example, without presuming that the example is complete from a software development point of view, let's compare two approaches in the form of an application workflow:

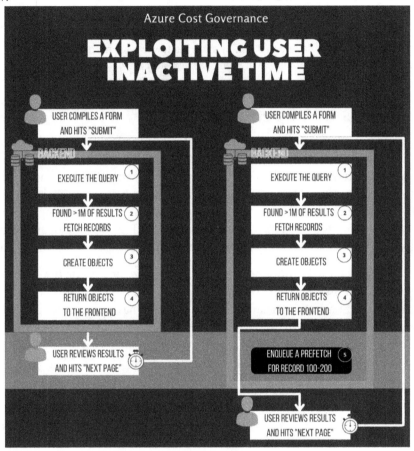

Figure 7.7 – Exploiting user inactive time

Usually, you're asked to finish the backend operation, return data to the frontend, and wait for the next user interaction. Here, the backend response time is directly related to the speed that the database and the application server can fetch records and create objects with.

What if, as in the workflow on the right, we take advantage of the time the user loses thinking about the results to prefetch other records? (Refer to *Step 5* in the preceding diagram.)

I have never found a user that needed to check results faster than an average *paged* backend query results elaboration, so we can use that time to give the illusion of a faster execution (since the results are already prefetched).

> **Important Note:**
>
> If the user skips three or four web pages, the illusion ends, and this trick won't work. We can mitigate this risk by showing fewer page numbers on the paginator on the frontend or leaving only the *next* and *previous* options.

Database sessions opening cost

Some applications or ORMs do not rely on a database connection pool (this maintains several open connections in the background and creates or recreates them based on the application's needs) but we need to open a connection every time the backend is called and close it before returning data and control to the user (or even more, they don't close it intentionally but let the connection idle timeout do the dirty work). Opening a connection each time the user submits an action/request costs time, and in this kind of application, it's always in the backend, so the user experience pays for that cost for every interaction.

What if we can modify the application to use a connection pool or recreate the connection using the seemingly lost user time? In both cases, we end up having a more contemporary (*modern* is not the right term) application and the user will have a *net* experience on the requested elaborations.

Profiling, profiling, and profiling!

At this point, you probably understand that you cannot know everything about your applications and your users' behavior, but unfortunately, if you want to optimize the infrastructure without impacting the user experience, you need to know as much as possible. Infrastructure (and cost) optimization is very limited without knowing about the application and intervening (if possible).

Fortunately, we have a couple of solutions:

- **Profiling the application**: All modern programming languages allow you to measure the timing of iterations, method invocations, object creations, and so on. Asking for the collaboration of application developers and maintainers to profile the application to identify bottlenecks and possible optimization (both infrastructural and architectural) is a mandatory step to achieve the best results.

- **Application Performance Management** (**APM**): You could instrument an APM to continuously monitor how the application is performing and identify bottleneck or optimization areas before they become critical for your users.

Learning about Azure Application Insights

At this point, it should be pretty clear that knowing your application is a key element of control, especially for cloud costs. **Azure Application Insights** is an APM tool that's available in Azure. It monitors many things and has native connectors for all the Azure services.

Having an APM tool is critical to understand if your application has substantial architectural issues, but also to spot that little hidden problem that might be found by one user after a million page hits. However, this will trigger a devastating number of faults.

For example, you may have followed all the modern cloud-native architectural patterns to refactor your application to be able to scale it up and down dynamically and even add one or two Azure Functions to simplify the integration. Once you fire it up, however, you soon find out that the performance of the Azure Function in the pay-as-you-go model is not optimal (you might consider tiering it up to a dedicated plan, which is faster but more expensive) and that the cost is way higher than keeping the same piece of code on a VM.

Functions as a Service (**FaaS**) is designed for short-lived, scalable code and you might find a different and cheaper way of achieving the same result. With Application Insights, you can see what your Azure Functions are doing and why, and then decide if they are a good fit for that specific piece of code.

Using Application Insights is quite easy as you have an SDK to add to your software. At the time of writing, the supported languages are C#|VB (.NET), Java, JavaScript, Node.js, and Python.

Embedded application cost optimization

If you kept reading and were intrigued by the reasoning we used in this chapter, one additional step might be to build something that will do most of the work for you. By knowing how your application works and what patterns move the cost needle up or down, you can think of writing a piece of code (I am thinking of a sidecar service with DAPR, for example) that will change some of the *expensive* patterns within your application once a specific threshold has been hit and you need to bring costs down.

This needs to be monitored together with the APM, as you probably don't want to sacrifice performance over savings (even though we'll have to talk about this later in the next chapter, anyway), but the final, ultimate cost governance process may be an application that is cost-aware and will initiate one or more cost-saving techniques when properly triggered.

With this final idea for an automated application optimization service, we've learned how our applications can use infrastructure resources (IaaS or PaaS) and how we can optimize them to consume fewer resources and, therefore, bring down costs. A very welcome side effect of this optimization is a consistent improvement in performance, though this might not necessarily be our primary goal.

Optimizing your database and application is not just a software development's best practice: in cloud, it can directly bring you extra savings from the over-allocated resources that you were running before the optimization and can help keeping your technical debt down. This should be a mandatory part of your overall Azure Cost Governance process.

Now, let's review everything we discussed in this section with some practical examples.

Practical examples

So far, we have analyzed the most common saving techniques, such as cleanup, right-sizing, and reservations, and have also considered and maybe already planned an application optimization continuous process. In this section, we'll bring everything together with some real-life examples that will show you how everything will come together in a beautiful dance once you've trained each *dancer* individually.

Storage throughput versus network bandwidth

Previously, we learned that every VM family has different capabilities and performance, so we have VMs with wider storage bandwidth and VMs with wider network bandwidth, with different costs. In addition, please always keep in mind that the network traffic inside a VNet is not billed.

Let's consider your first application: it's a one-database architecture, where it is unlikely that you can save costs by reducing nodes as they are already at a minimum. Decoupling the database storage to another VM might even add costs instead of saving, as shown in the following diagram:

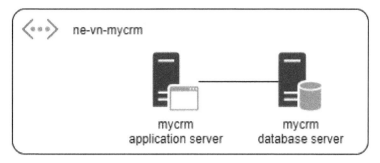

Figure 7.8 – Sample architecture with one database

But your applications might not constantly access the database, so you can share a network storage solution between database servers, trying not to exceed the storage overbooking.

> **Important Note:**
>
> Remember to always monitor the throttling (the capped capacity) on disks and networks when sharing resources, thereby satisfying multiple compute/ RAM needs with a single VM. If you spot a sustained plateau in the graphs, and if it corresponds with an underperforming database, that is most certainly a symptom of under-provisioning.

Let's analyze an example of the database usage pattern – four databases are used differently during the day, all residing on an IaaS VM:

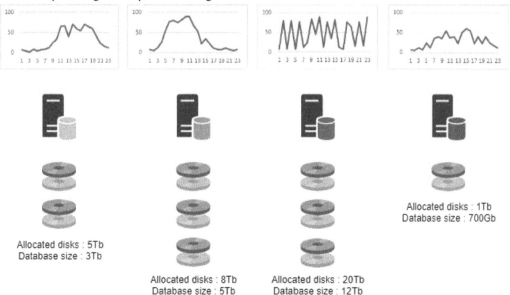

Figure 7.9 – Disk allocation and the related application traffic

As you can see, we have a heavy load during working hours, but during the night, we have an impulsive and low but continuous load. We need a lot of disks to sustain the peak performance demand. These are also attached during non-peak hours, and since this may be a critical application, you cannot switch it off and rescale disk tiers, as we learned in *Chapter 4, Planning for Cost Savings – Right Sizing.*

Let's assume that by analyzing the CPU load percentages, we find that 12% to 17% is **iowait** on the disks. At this point, we have gathered the following information:

- We have evidence that the four workloads can be consolidated on a single VM.
- Ensuring we have enough RAM, we can take advantage of database caching (relieving the database from continuously accessing the disks and leaving the remaining available IOPS/throughput to other shared loads).
- Traffic to/from different subnets (within a VNET) is free.
- Decoupling only one database's host storage is not cost-effective.
- Some VM sizes are network-oriented, while others are disk-oriented.
- You may need bigger VMs than what you expect, just to attach more disks.
- PaaS is always an option, as we'll explore later in this section.

At this point, our idea is to use an IaaS **network-attached storage** (**NAS**) VM, as we'll see in the next section.

IaaS-to-IaaS example

We can try to evolve our IaaS infrastructure a bit by adding a NAS VM that shares all the storage via a network with the database servers. This approach will consolidate all the database storage disks into volumes that have been shared from an IaaS NAS.

The following diagram summarizes the concept of still having four database servers (right-sized to a lower VM family), migrating the database storage to a single NAS, reducing the total amount of allocated disks, and minimizing the free space and the wasted space:

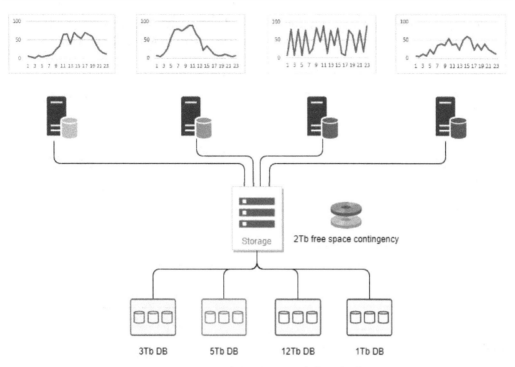

Figure 7.10 – A single NAS VM with four databases

With this new approach, we understand the following:

- We saved about 10 TB of allocated disks because we size the *striping* based on the max throughput, but the space (and throughput) is shared between all the databases.

- Sharing the disk space will reduce the free storage waste too.

- We will need to change the database server's VM family from storage-oriented to network-oriented.

- We can right-size the database servers because we won't have a high capped capacity on disk throughput, thus saving on compute.

Is there anything else we can do to optimize this application? We'll look at this in the next example, where we'll be swapping the IaaS VM for a PaaS service.

IaaS-to-PaaS example

This approach will consolidate all the database storage disks into volumes that have been shared from an Azure NetApp Files volume.

The following diagram summarizes the concept of still having four database servers (right-sized to a lower VM family) and migrating the database storage to Azure NetApp Files:

Figure 7.11 – Four database servers with one Azure NetApp Files volume

With this layout, we understand the following:

- We saved more than 10 TB of allocated disks because we size the *striping* based on the max throughput, but the space (and throughput) is shared between all the databases.

- Using PaaS storage (with a pay-per-use approach) allows us to save on free space for immediate contingency too because Azure will provide storage when needed, without pre-allocating it.

> **Important Note:**
> Opposite to the IaaS-on-IaaS approach, we don't have managed disks that are manually allocated to the NAS, so it takes nearly zero time to hot-extend the volume and have more space.

- We will need to change the database server's VM family from storage-oriented to network-oriented.

- We can right-size the database servers because we won't have a high cap on disk throughput, thus saving on compute.

- In the case of throughput or IOPS increases, since the storage is now using a PaaS service, you can use AZ commands or APIs to *hot-reconfigure* the volumes, thus *shaping* the volume performance to your needs – even in a dynamic way – by following the performance of the application.

But wait – can I do something smart on the database side as well? We'll find out in the next section.

Iaas to PaaS[2]

This approach relies on having the four databases available in PaaS from Azure as well (for example, they could be MariaDB, MySQL, PostgreSQL, or SQL Server). Naturally, if the source database is not supported on Azure, you can't take advantage of this approach directly, so you might need to do some rework to migrate to a supported PaaS database.

The following diagram summarizes the concept of migrating four databases (**DB**) from IaaS to PaaS (*the size of the database icons refers to the allocated storage space*):

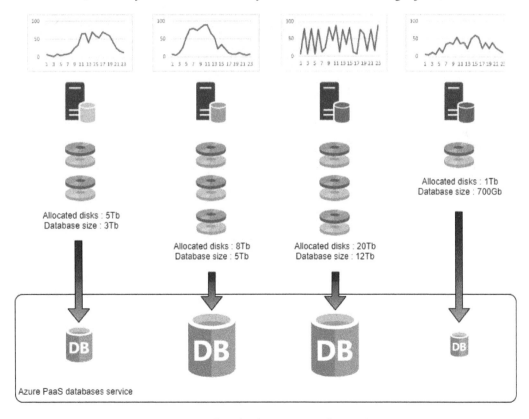

Figure 7.12 – Our four databases migrated to PaaS services

The migration can be as simple as a backup/restore, but you need to consider that the following:

- Applications may need to be reconfigured to connect to the new database endpoint.

- If you want the public traffic from the application servers to the PaaS databases to be *vehiculated* directly on your virtual data center via the Azure infrastructure, you need to check and enable service endpoints in the VNet configuration of the application servers. Here is a list of available endpoints at the time of writing: `https://docs.microsoft.com/en-us/azure/virtual-network/virtual-network-service-endpoints-overview`.

- I suggest evaluating whether you leave the public database access open (with appropriate identification and authorization) or set up private IPs for the PaaS database servers, thus shutting down public access.

With this approach, we understand the following:

- We can now switch from a PaaS database tier to another, with zero or little downtime (which is typically related to the size of the database – the bigger the database, the longer the downtime).

- We only pay for the used storage and the computational power, depending on the choice of database and its pricing.

> **Important Note:**
> Many PaaS databases already incorporate some *elastic scale* functionality (with just a flag enabled!), but if this isn't the case, you can create automation for your databases to resize them before and after the peak, according to our well-known demand shaping technique.

A basic monolith app

We've learned that among the many advantages of cloud-native microservices applications, scaling is the most important cost-wise because it allows us to tailor the number of services to the necessary resources, just as we want it to be. *But what if your application is a monolith? Can you still find ways to scale it and save on under-utilized resources?* There are several options to play with, so let's review them in the following practical example.

Azure VMSS example

Azure **Virtual Machine Scale Set** (**VMSS**) is a service that helps you set up a group of load-balanced VMs that can start either from a standard image or, as in our example, from a custom image containing all your application's software. The beauty of a VMSS is the scaling, which can be either manual or automatic, and will create high availability for your application by increasing and decreasing the nodes behind the load balancer.

To move your application to a VMSS, the application must support a load balancer that will direct the traffic toward the scaled nodes. In the following steps, we'll learn how to generalize our VM, create an image, and then create a VMSS for it:

1. First, let's create a resource group for our tutorial:

    ```
    az group create --name vmsstest --location westeurope
    ```

2. Now, we need to create a custom VM image with our application. This will be used to spin up all the nodes of the scale set when needed. To create a custom image, the VM must be deprovisioned, deallocated (stopped), and used to create the new image.

    I am using a Linux VM for this example. On your Linux VM, run the following code:

    ```
    sudo waagent -deprovision+user
    ```

    On the Azure CLI, execute the following commands:

    ```
    az vm deallocate --resource-group vmsstest --name
        sourceVM
    az vm generalize --resource-group vmsstest --name
        sourceVM
    az image create --resource-group vmsstest --name
        cimage --source sourceVM
    ```

    Please note that the last command depends on the resource being created and will fail if this is missing.

3. Create a new VM from the image to test it. Check that the application works as intended. Once we have our custom image, we can create an image definition:

    ```
    az sig image-definition create --resource-group
        img_RG \
        --gallery-name my_imgs \
        --gallery-image-definition my_img_def \
    ```

```
--publisher my_img_pub \
--offer myOffer \
--sku mySKU \
--os-type Linux \
--os-state specialized
```

4. Now, we need to create the image version. Please remember to use your own subscription ID in the following code:

```
az sig image-version create \
    --resource-group img_RG \
    --gallery-name my_imgs \
    --gallery-image-definition my_img_def \
    --gallery-image-version 1.0.0 \
    --target-regions "westeurope=1" "eastus=1" \
    --managed-image "/subscriptions/<Subscription
        ID>/resourceGroups/MyResourceGroup/providers
        /Microsoft.Compute/virtualMachines/sourceVM"
```

5. Then, we can create our VMSS from the custom image:

```
az vmss create \
    --resource-group vmsstest \
    --name vmss_test \
    --image "/subscriptions/<Subscription
        ID>/resourceGroups/ img_RG/providers/Microsoft
        .Compute/galleries/my_imgs/images/my_img_def" \
    --specialized
```

Once the VMSS has been created, we can test it and enable autoscaling from the portal page. You can also create a template to automate this using the documentation at `https://docs.microsoft.com/en-us/azure/virtual-machine-scale-sets/virtual-machine-scale-sets-mvss-start`.

Once we have put our application on a VMSS, we can benefit from all the features and power of its scaling, though we might want to try a different way to make autoscaling much faster and enable solid orchestration. A possible choice is to containerize the application, as we'll explore in the next section.

Exploring containers

Another common way of improving your monolith application is to containerize it. The benefits of containerization are immediate: you can define automation for autoscaling and provisioning, you can move it to the cloud, even if you are not planning to touch it, you can instantly replicate it for dev and test, and finally, you can scale it up and down according to its traffic, releasing unused resources and taking advantage of the cloud paradigm.

Even if the application is not strictly monolithic, the result will be a single container – a single process that contains everything the legacy application had. Compared to the VMSS option, this can be a much cheaper solution, as many containers can run on a single VM. However, if you are not familiar with Kubernetes, there might be a learning curve in terms of managing containers.

If you can refactor the application by splitting the monolith into more services, this could even bring extra optimization to the table, as the containerized services might be able to scale independently without requiring a full scale. A typical example is a web commerce application with a frontend and backend ordering system: you might have tons of hits on the frontend but only a percentage will reach the ordering path, and if you can split the frontend and backend into separate containers, you can scale them according to their real usage.

Containerizing an application is quite easy and depending on your choice of deployment, you can find lots of tutorials and documentation online to guide you through this work.

> **Important Note:**
> As we have often noted, Azure offers ready-to-use tools to simplify your starting point. For containerization, the new App Containerization tool from Azure Migrate is our recommended tool: it will automatically discover and build a container image, and then deploy it to Azure Container Registry and Azure Kubernetes Service in just a few steps. More information can be found in the official documentation: `https://docs.microsoft.com/en-us/azure/migrate/tutorial-app-containerization-java-kubernetes`.

The main steps, which include starting with a VM that contains all your executables and libraries, are summarized in the following diagram:

Figure 7.13 – Containerizing a monolith application

The steps range from preparing your base application image and `docker-compose` file to running the container (checking whether everything works fine). Then, we can upload the image to a registry such as **Azure Container Registry** and use a service to run it, such as **Azure Kubernetes Service**.

Using serverless resources

Several times throughout this book, we have mentioned that if you plan on consuming just the right resources, then PaaS services are your best bet and that, being the most optimized cloud-native service set, serverless computing is the way forward. Serverless compute on Azure has several implementations:

- **Function as a Service (FaaS)**, with Azure Functions, lets you run your portion of code in a lambda function object.

- **Logic Apps,** with its easy and powerful graphical programming flow and tons of connectors.

- **Serverless Kubernetes**, with many choices of deployment, via **Azure Kubernetes Service** (**AKS**) with the *unlimited containers* virtual pod to the App Service hosting, to **Azure Container Instances** (which is in its early stages and preview at the time of writing, so it's a great stepping stone toward AKS).

- The newly launched **Azure Container Apps** (in preview, at the time of writing), which incorporates DAPR and can scale dynamically through **Kubernetes Event-Driven Autoscaling** (**KEDA**) (`https://keda.sh/`).

When you're considering your monolith application, the most common approach to a serverless architecture is the so-called **Strangler pattern** (from the original design pattern by Martin Fowler, where you start splitting off services from your application and implementing them one by one by following a serverless approach):

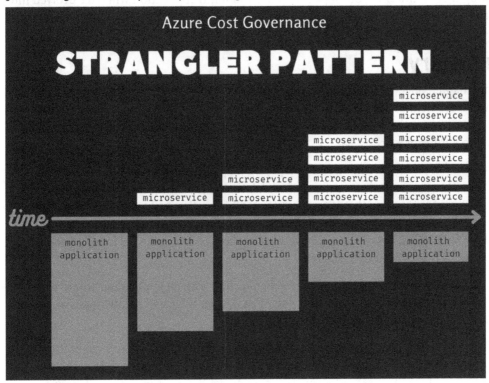

Figure 7.14 – The Strangler pattern for migrating a monolith application

There are several benefits to this approach:

- The refactoring is minimal as it will only impact all the areas in the application that are using the service you are replacing.

- The application has no major downtime as the new service can be run in parallel to test it with different deployment approaches (we are not going into this now, but you can go to the *Further reading* section to find out more).

- The legacy code of the moved service can be removed, and the code gets leaner with every iteration.

- The newly created microservice, using serverless, is now the most scalable, efficient, and cost-effective bit of your legacy application, so you may start saving money immediately if you choose what service to refactor wisely.

Azure offers many serverless choices to provide an ideal platform for this type of approach, and the greatest benefit is that you can work on this pattern at your own pace, knowing that, in the end, your *ugly caterpillar* monolith application will have evolved into a beautiful modern, cloud-native butterfly.

Summary

In this chapter, once we had covered every possible cost-saving implementation in terms of the infrastructure, we started analyzing our application catalog to understand how and why optimizing at the application level is sometimes the only way of bringing extra cost saving to your process.

We started with database tuning and optimization by providing an optimization plan template. Then, we looked at what technical debt is and how to understand, analyze, represent, and reduce it so that your application set is continuously enhanced and improved by keeping costs down. We discussed the importance of an APM tool, such as Azure Application Insights, not only to manage your application at a technical level but also to spot issues that may incur unexpected costs.

We then worked on a few practical examples, starting with an application where we moved the database to NAS (first IaaS, then PaaS), then migrated the database to PaaS, to increase efficiency and cost-effectiveness.

The last part of this chapter was dedicated to a very common situation: a monolith application that we can initially move to a VMSS, then to a container, then service by service, and then to a serverless architecture, to get the most out of cloud-native application objects.

The next and last chapter will link everything we learned so far to a topic that is most dear to me – sustainability, and how we can incorporate it as a form of motivation to improve our software.

Questions

1. What are the limitations of infrastructure-only optimization, and why do you need to work with the app and dev team for any application optimization?

2. Why can database tuning be important for cost-saving?

3. What is technical debt?

4. Why is Application Insights a key element of application optimization?

Further reading

*   Azure Application Insights: `https://docs.microsoft.com/en-us/azure/azure-monitor/app/app-insights-overview`

*   Azure application architecture fundamentals: `https://docs.microsoft.com/en-us/azure/architecture/guide/`

*   Benefits of using Azure NetApp Files with Oracle Database: `https://docs.microsoft.com/en-us/azure/azure-netapp-files/solutions-benefits-azure-netapp-files-oracle-database`

*   Strangler Fig Pattern: `https://martinfowler.com/bliki/StranglerFigApplication.html`

8

Sustainable Applications and Architectural Patterns

In the previous chapter, we explored and learned how optimizing your applications has a *positive* impact on your Azure services' costs.

This chapter will focus on sustainable software engineering, how an application which is designed for performance and its streamlined cloud cost can also have a lower impact on its carbon footprint.

In this chapter, we'll cover the following topics:

- Understanding sustainable software
- Demand shaping and shifting
- Sustainable cloud-native architectures
- Measuring and fostering sustainability

Upon completion of this chapter, you will have a good understanding of the most recent, performing, and sustainable architectural patterns for a low-cost and carbon-aware application.

Technical requirements

For this chapter, you'll need the following:

- A computer with internet connectivity

- Access to the Azure portal

- Privileges to access Cost Management information (see `https://docs.microsoft.com/en-us/azure/cost-management-billing/costs/assign-access-acm-data`)

Understanding sustainable software

In this section, we'll learn about the new discipline of **sustainable software**, also known as **green software engineering**, and how it can be used to design applications that can save on carbon emissions. Carbon-saving techniques are also related to cost governance and application optimization, and in some cases, we will learn how the tips and tricks for cost saving that we used previously can also be utilized for building green applications.

Defining a need for sustainability

The whole scientific community has been agreeing for some time now that the world has a pollution problem. The **carbon dioxide (CO2)** in our atmosphere has created a layer of gas that traps heat and changes the earth's climate. Earth's temperature has risen by more than 1 °C since the industrial revolution of the 1700s. If we don't stop this process of global warming, scientists have warned us that the consequences will be catastrophic, resulting in the following:

- Further increase in temperature.

- Extreme weather conditions such as droughts and fires (*Do you remember the Australian situation at the beginning of 2020?*).

- Rising waters could make areas, where more than 200 million people live today, uninhabitable.

- Drought will ultimately lead to food shortages, which could impact over 1 billion people.

To sum it up: we need to *drastically reduce CO2 emissions* and prevent the temperature from rising beyond 1.5 °C.

But here's another problem: each year, the world produces and releases more than 50 billion greenhouse gases into the atmosphere. Unfortunately, society and politicians are still struggling to take this seriously; there is still little commitment to a valid plan. The first step toward understanding the problem is to analyze a bit of *carbon math*. These are the basic concepts that help us understand how the problem of pollution applies to all of us, as individuals, families, professionals, companies, or other organizations.

CO2 emissions are classified according to three main categories, as follows:

- **Scope 1**: Direct emissions created by our activities.

- **Scope 2**: Indirect emissions that come from the production of electricity or heat, such as traditional energy sources that power and heat our homes or company offices.

- **Scope 3**: Indirect emissions from all other daily activities. For a company, these sources are many and must include the entire supply chain, the materials used, the travel of its employees, the entire production cycle, and more.

We commonly speak of *carbon efficiency*, even if greenhouse gases are not only made up of CO2 and they do not all have the same impact on the environment. For example, 1 ton of methane has the same heating effect as 80 tons of CO2, so the convention used is to normalize everything to the *CO2-equivalent* measure and call it **carbon**. International climate agreements have ratified the reduction of carbon pollution and agreed to stabilize the temperature at an increase of 1.5 °C by the year 2100.

And here's yet another problem: the temperature increase does not depend on the rate at which we emit CO2 but on the *total amount present in the atmosphere*. To stop the temperature increase, we must therefore avoid adding to the CO2 already present—or, as they say, reaching the goal of *zero emissions*. Of course, to continue living on earth, this means that for every gram of CO2 emitted, we must take the same amount out. The goal is ambitious; I don't mean it's impossible, but we must commit to it immediately: we need to reduce emissions by 45% by 2030 and achieve zero emissions by 2050.

But isn't this is an Azure Cost Management book? So, let's shift our attention to our public cloud world where we are directly involved and have a look at what's happening there, as follows:

- The demand for compute power is growing faster than ever, thanks to the innovation of public cloud offerings, including Azure.

- Some estimates indicate that data center power consumption will account for no less than one-fifth of global electricity by 2025.

- A physical server or a **virtual machine** (**VM**) on average operates at 20 to 25% of its processing capacity while consuming a lot of energy.

- On the other hand, in an instance where applications are running using physical hardware, it is still necessary to keep servers running and use resources, regardless of whether an application is processing or not.

- The same is true for a public cloud data center: once it has been turned on, the only way of using it efficiently is to saturate all the resources. If only this were possible.

- Physical (single-server) hardware and its utilization have improved with hardware virtualization (VMware, Hyper-V, and more), along with a reduction of the related emissions.

- *Containerization* is **operating system** (**OS**) virtualization layered on top of the already virtualized hardware, making it even more efficient: containers have a higher density and can bring a server up to 60 to 80% of compute capacity utilization. This is good but we can do better, especially since not all applications run on containers.

- Ultimately, 40 to 60% of the world's server capacity is estimated to be idle at some point.

> **Important Note:**
>
> Data centers, and especially high-scale and highly optimized public ones such as Azure, have been working for several years on **power usage effectiveness** (**PUE**), which is the ratio between the amount of energy used and the energy delivered to servers, mostly for cost concerns. Currently, every cloud provider is also working on sustainability; so, this means that once our infrastructure is optimized and cost-efficient, there is very little that we need to do to improve it. This means that we need to focus on the application's efficiency.

We have two ways of improving carbon efficiency in the Azure cloud: by improving the efficiency of the infrastructure and/or improving the efficiency at the application level. *Does this ring a bell?*

Let's dive into how we can make this a discipline for sustainability as much as for cost governance.

Principles of sustainable software engineering

The starting point for green software is the `principles.green` website, where a community of developers and advocates started creating guidelines for writing sustainable code a few months back. This eventually merged in a much bigger operation, which at the time of writing is called the **Green Software Foundation** (**GSF**) (`https://greensoftware.foundation/`), launched in May 2021.

The goal of this foundation is to create standards and measurements, and the basis for a movement that will bring sustainability into the hands of developers who can make daily choices about their applications. The main principles of sustainable software engineering that will be enriched and integrated with new ideas and methodologies are listed here:

- **Carbon**: The first step toward sustainability is to have *sustainability of an application* as a general target. It seems trivial, but even today, there is not much documentation about it in computer textbooks or websites.

- **Electricity**: Most electricity is produced from fossil fuels and is responsible for half of the carbon emitted into the atmosphere. All software requires electricity to run, from an app on a smartphone to the **machine learning** (**ML**) models that run in a cloud data center. Developers generally don't have to worry about these things: electricity consumption is usually referred to as *someone else's problem*. But a sustainable application must take charge of the electricity consumed and be designed to consume as little as possible.

- **Carbon intensity**: Carbon intensity is a measure of how much carbon is emitted per the electricity consumed. Electricity is produced from a variety of sources, with different emissions in different places and at different times of the day, and most of all, when it is produced in excess, we have no way of storing it (yet).

- **Embedded or embodied carbon**: This is the amount of pollution emitted during the creation and disposal of a device. Efficient applications that run on older hardware also have a good impact on emissions.

- **Energy proportionality**: The maximum rate of server utilization must always be the primary objective. In general, in the public cloud, this also often equates to cost optimization. The most efficient approach is to run an application on as few servers as possible and with the highest utilization rate.

- **Networking**: Reducing the amount of data, and the distance it travels across the network, also has its impact on the environment. Optimizing the route of network packages is as important as reducing the use of the servers. Networking emissions depend on many variables: the distance crossed, the number of hops between network devices, the efficiency of the devices, and the carbon intensity of the region where and when the data is transmitted.

- **Demand shifting and demand shaping**: Instead of designing demand-based applications, green software will get the demand from the energy supply. Demand shifting involves moving workloads to regions, and at times with lower carbon intensity. *I even found a Carbon Kubernetes scheduler that will move pods according to the carbon impact of the region!*

  Demand shaping, on the other hand, involves separating workloads so that they are independently scalable and prioritizing them to support features based on energy consumption. When the energy supply is low, the carbon intensity is higher, and so the application reduces its features to a minimum, keeping it just at the essential capabilities. Users can also be involved in the choice by presenting them a *green* option with a minimum feature set.

- **Monitoring and optimization**: Energy efficiency must be measured in all parts of the application to understand how to optimize it. *Does it make sense to spend 2 weeks reducing network communication by a few megabytes (MB) when a full-scan database query has 10 times the impact on emissions?*

How can developers have an impact on carbon? Here are a few ways:

- By making a program more accessible to older computers and devices.

- By writing code that exchanges less data, has a better **user experience** (**UX**), and is greener.

- By co-locating tightly coupled microservices when needed, to reduce network congestion and latency.

- By considering running resource-intensive microservices in a region with less carbon intensity.

- By optimizing the database and how data is stored, reducing the energy to run it and therefore reducing all the idle times and pending queries.

- In many cases, web applications are designed by default with very low latency expectations: a response to a request should occur immediately or as soon as possible. But this may not be necessary after all. Evaluating whether latency limits can be eased in some areas can also help reduce emissions.

Now that we know the starting point (which is the need for an environmental change) and the weight of data centers in the carbon picture (through the principles of a sustainable application), let's wrap up what we have learned in the previous chapters of this book and apply it to a sustainable software strategy.

Demand shaping and shifting

We have learned that one of the principles of sustainable software engineering is **demand shifting** and **demand shaping**. *Haven't we seen this before?* Yes—we learned how to perform demand shaping from a cost-perspective point of view in *Chapter 4, Planning for Cost Savings – Right Sizing*, under the *Sample logic for cost control* section; so, let's see if this can also be applied to carbon efficiency.

Demand shaping

Demand shaping can and should be applied to change the culture and perception that end users have toward applications. For example, when we publish a website with lots of graphics and images, we could offer a lighter version of the website or interface to end users by providing a *green* **Button** option (see *Figure 8.1*) that will allow energy (and carbon) saving.

Another common option is to target the top 2 to 5 most used features and package them as the *basic green version*, leaving the full version to the choice of users. This has an overhead in the code, obviously, but will start changing the culture of our current digital end users, who are completely relying on massive streaming of video and pictures without wondering what impact this *waste* of bandwidth has on the planet.

In the following screenshot, an example of a **GO GREEN!** button that users can choose to have a simplified but sustainable version of the same app is shown:

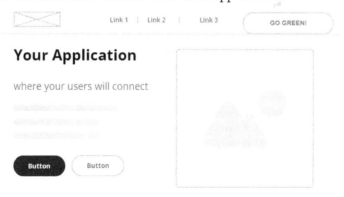

Figure 8.1 – An application mockup showing the go green button

We, as developers and application owners and maintainers, should start involving end users in this decision: first, by measuring what an application's carbon impact is, then by providing green choices to dynamically improve it. This will have a positive side effect of fostering a different digital culture—one that doesn't revolve around faster performance and higher bandwidth requirements, but rather on more frugal choices that tip the scale toward a greener and less carbon consuming digital experience. Let's now see what demand shifting is.

Demand shifting

In addition to creating a culture of green software usage, **demand shifting** can also be easily achieved thanks to our orchestrator's exercise seen in *Chapter 4, Planning for Cost Savings – Right Sizing*. Once you are able to measure the cost, you can move your analysis to the carbon impact of a specific service, site, and moment in time so that you can dynamically decide to move your workloads (where possible) to a site with a lower impact.

As a starting point, whenever you are lowering an application's usage for cost purposes, you are in fact also lowering the carbon impact of that workload. Soon, Azure will provide carbon-impact information about its services, which can then help us create carbon-aware applications that can actively shift loads toward less impacting sites.

In this section, we've learned how to reduce the scale of an application's carbon impact. In the next section, we'll talk about architectures that are inherently greener, and this should be your first choice when targeting a sustainable application.

Sustainable cloud-native architectures

Cloud-native architectures are those that leverage the Azure public cloud as a developers' platform and not just another infrastructure data center. They have many advantages that we have mentioned throughout this book: great scalability, resiliency, better performance, and the flexibility we are used to in the cloud. In addition, if they are done right, compared to legacy **infrastructure-as-a-service** (**IaaS**) or on-premises apps, they are generally cheaper. One of the latest trends of cloud-native apps is using **serverless** services, which have many additional benefits on top of being cloud-native, including the following:

- From an infrastructural point of view, the use of serverless allows for more efficient use of the underlying servers, precisely because they are managed in shared mode by the cloud suppliers and built for an efficient use of energy to obtain optimal data center use.

- Cloud data centers have stringent rules and often, as in the case of Microsoft, have ambitious targets in relation to emissions (Microsoft recently declared its will as a company to become carbon negative by 2030: `https://blogs.microsoft.com/blog/2020/01/16/microsoft-will-be-carbon-negative-by-2030/`). Making the most of the already optimized resources of a cloud provider implicitly means optimizing the emissions of your application.

- Serverless and stateless workloads, as event-driven architectures, are often ready for demand-shifting/-shaping executions, thus allowing us to obtain a dynamically eco-sustainable architecture with a few tricks, without too many configurations.

- From a purely theoretical point of view, writing optimized and efficient code is always a good rule, regardless of the purpose for which it is done.

- These architectures foster a simplicity of design that will make your technical debt management easier.

- Their scalability is fully automated, and you can pay as you consume. You even save the effort of building orchestrators and schedulers!

Serverless architectures are highly optimized and efficient, both from the point of view of the vendor (the data center that hosts them) and from that of the customer who uses them. Users obviously have an eye on the costs and therefore tend not to waste useless calls to **Functions** and **Logic Apps**, thus they represent the arrival point for cloud-native and green applications.

Let's see a practical example of a green serverless application, in the next section.

Example of a green serverless app

Imagine you are managing the following application: you have users taking pictures and sending them to a global storage. Depending on the usage of those pictures, you explore your requirements and find out your needs, which comprise the following:

- A lower quality

- A subset of the picture (for example, recognizing objects)

- Text extracted from the picture (such as reading a metering value)

Do you need to keep the full quality picture after you extract the needed information? If the answer is no, then you might consider resizing it to a lower resolution, thus lowering its carbon footprint. In addition, you can do so with Webhooks, Azure Functions, and Azure Blob Storage, without the need to allocate specific resources.

Relying on the serverless pattern will allow you to scale as much as you need, paying for just the usage duration of the Lambda function. The result, as you can see in the following screenshot, is a very simple, clean, and efficient architecture that follows the principles of sustainable software engineering, in terms of resource usage. If we were to compare the usage of this pattern versus the consumption that would be required on a physical server or a VM, we'd be targeting a lot of *wasted* emissions, and if every company were to do this with a highly optimized public cloud such as Azure rather than in their own data center, then the carbon impact of this seamless small change would definitely be staggering:

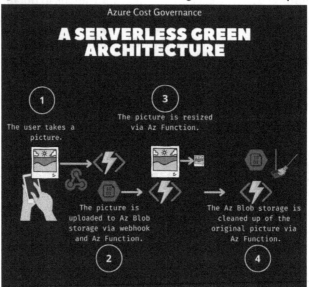

Figure 8.2 – Our serverless green architecture

Could we further improve the carbon efficiency of this application? Well, for example, if we consider the broader scope up to the end-user device, we should analyze the carbon cost of resizing the picture locally at a device level so that we don't waste bandwidth, and then evaluate whether this saving is comparable to the saving when centralizing this resizing operation.

This is still an ongoing discipline, and I am sure that the Azure team will soon come up with carbon-aware services that will allow you to embed these considerations directly into your native applications!

My conclusion, though, is that serverless (if properly used) is the future of cloud-native applications, not just because it provides beautiful, practical, and affordable architectures, but also because it has the least impact on emissions to date.

Carbon optimization of an on-premises application

In the previous sections, we learned some key points on how to modernize an application and migrate it to a greener architecture. Now, let's try to apply the concepts to a practical example.

Let's start with our on-premises legacy application and follow its path toward a modernized, more sustainable architecture. The following diagram illustrates a typical legacy architecture of an on-premises application, equipped with a firewall, reverse proxy, load balancer, frontend, backend, batch, queue, and several database nodes, all in redundant or cluster availability:

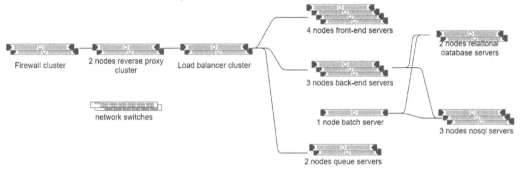

Figure 8.3 – High-level diagram of an on-premises architecture

The infrastructural starting point can be something like this, with **high availability (HA)** (either physical or virtual):

- Four network appliances
- Two reverse proxy servers
- Four frontend servers
- Three backend servers
- Two servers for queue services
- One batch server
- Three servers for a NoSQL database
- Two servers for a relational database cluster

Of course, on top of it all, we have networking, storage, and facilities services, and the cooling system too.

Everything must run 24/7: usually, in a data center, you don't power on or off the servers or VMs but always leave them on, especially when in production.

Even if it's always powered on, it's quite difficult to estimate the carbon impact of such architecture because of the device's power, cooling facilities, consumption, and more.

Lift and shift cloud migration

Usually, **lift and shift migration** is the first step to the cloud, since it doesn't require major changes to the application: we can start maintaining a 1:1 ratio for the **virtual central processing unit** (**vCPU**) and **random-access memory** (**RAM**), speeding up the migration process. In terms of migrating your application to Azure, Microsoft states that (at the time of writing) by 2025, every device or server that hosts your services will be powered by renewable energy.

Furthermore, thanks to the Microsoft Emissions Impact Dashboard, you'll be able to measure and constantly monitor the impact of your application and be able to measure the effect of technical (refactoring and modernization) or business (that is, a change in traffic or data volume) evolutions.

Please consider that during a lift and shift migration to the cloud, you don't need switches and routers anymore, because the virtual network/subnets and routing policies are Azure services that run on shared and already optimized devices.

Paying only for running instances

After migrating to the cloud, we switch from a *pay-once* to a *pay-per-use* paradigm. This should lead us to apply all the cost-saving techniques we learned in the previous chapters—for example, the following:

- Right-sizing instances and disks
- Powering off unutilized instances when not needed (maybe during the night, depending on your workload usage)

Minimizing the uptime of resources, using slower disks, and lowering the total amount of VCPUs/RAM used by your application will consequently lower the carbon impact.

In the following diagram, a *small clock* appears on all the components that, in our example, can be powered off during non-peak hours:

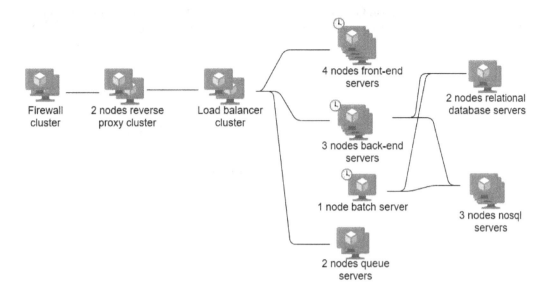

Figure 8.4 – Application migrated to Azure, with off-peak hours powered-off VM

Initially, after implementing basic cost saving techniques, you'll have a positive impact on your carbon emissions too, and you can monitor them using the following:

- Azure Cost Management
- Microsoft Emissions Impact Dashboard
- Microsoft Sustainability Calculator

The need for network appliances

With the lift and shift migration, we also might have migrated our network appliances such as firewalls or load balancers. *But do we need IaaS virtual appliances?*

Can they be replaced with native services such as Azure Firewall or Azure Load Balancer?

Again, the reverse proxy servers usually exist in a **DMZ** to analyze traffic, maybe apply some redirect or local caching, and forward the request to the backend firewall/load balancers. *Do we need two VMs running as a reverse proxy, or can we implement the rules on an Azure Application Gateway?*

If you chose Azure Application Gateway, then we may land on the infrastructure shown in the following diagram, reducing the carbon and cost impact of six devices:

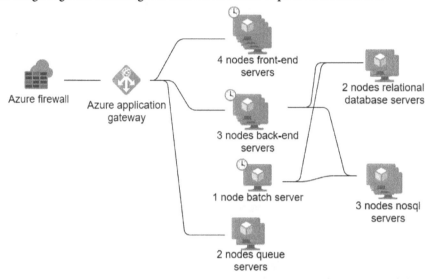

Figure 8.5 – Firewall, proxies, and load balancer replaced with Azure native services

We have replaced two Azure VMs (lifted and shifted firewall virtual appliances) with Azure's native firewall service. We then condensed the two VMs for the reverse proxy cluster and another two VMs (load balancer virtual appliances) with an Azure native component application gateway.

Lazy PaaS adoption

Let's consider the world of Azure **platform-as-a-service** (**PaaS**) services: these are energy-efficient services tools, shared by definition, that are available for your applications. They can be a great step in reducing your carbon impact and give you the flexibility to change their service performance and tier without (or nearly without) any downtime.

Unfortunately, in my experience, switching from legacy applications to PaaS services takes time and should be carefully planned (data migration, endpoint and protocol compatibility, service versions, and more). So, one approach is to concentrate on a small subset of quick wins until every possible service has been switched to PaaS, starting from the following:

- More compatible PaaS services
- More easily migratable services

Let's assume that, at first, we identify the queue servers as migration candidates to the Azure event hub, and this comes with a low impact on the application's code. We could then update the application map introducing the event hub, removing two VMs. The following diagram then shows how our queue servers have been replaced with a much more scalable and easier-to-manage event hub:

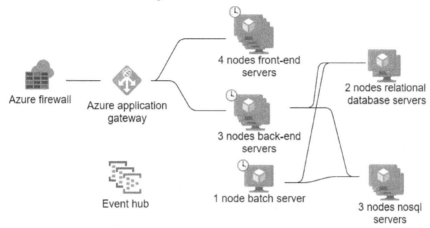

Figure 8.6 – Queue servers replaced with an event hub

After testing and running the application for some time, we can then measure the carbon impact of our choice and eventually carry out another analysis to migrate something else on a PaaS service.

Let's assume that we then decide to switch the NoSQL servers to Cosmos DB (for example, using the MongoDB-compatible interface), our application architecture will therefore change to resemble the following diagram:

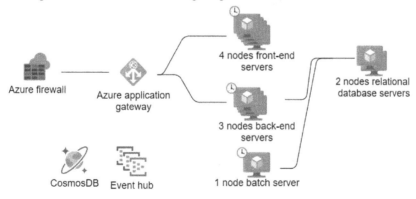

Figure 8.7 – NoSQL servers replaced with Cosmos DB Azure native service

For the last conversion, we tackle the relational database: it has lots of data to be migrated and its performance should be carefully tested. The evolution of our application now looks like this:

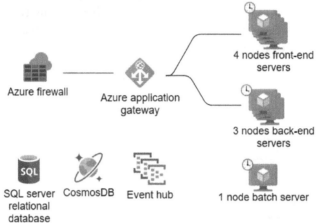

Figure 8.8 – Relational database replaced with Azure PaaS service (SQL Server)

We then realize that some smaller components might be switched to serverless services. So, let's start with recoding the batch server procedures to be executed as Azure Functions. And now, the layout looks like the one shown in the following diagram:

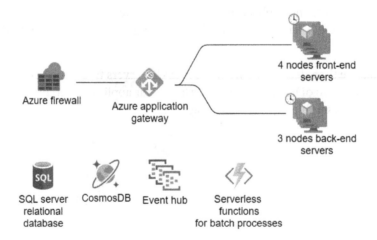

Figure 8.9 – Batch server/processes replaced with Azure serverless functions

The last component we can address is the frontend layers, as we'll learn in the next section.

The GUI component

We've now successfully migrated to PaaS with lots of services that were originally implemented (or just migrated) within IaaS components, thus reducing the overall carbon footprint of our application.

Let's now go a little further on the exercise: we'll analyze the frontend layer to convert it to an Azure web app (with a bit of refactoring). This step will remove the scheduled on-off for the frontend servers and enable the frontend layer to be automatically scaled based on **HTTP** or load metrics, introducing even further savings, as shown in the following diagram:

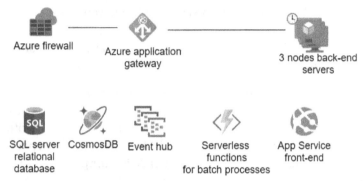

Figure 8.10 – Frontend servers replaced with App Service Azure native service

Let's assume that we cannot convert the backend servers to App Service. There is no other possible refactoring from an infrastructure point of view, but overall, we have evolved about 90% of our application from IaaS to PaaS.

In the previous sections, we learned how to tackle a progressive migration from IaaS to PaaS for our applications once migrated to the cloud, with positive effects on sustainability.

The first assumption was that the Azure infrastructure is highly carbon-optimized, and we now have an almost complete cloud-native application. *Are we sure we've done everything we can to reduce its carbon impact?* Let's explore this in the next section.

Programming languages and sustainability

Let's switch to the application layer, and—more specifically—to how the application is implemented (queries, internal flows, checks, cycles, cleanness, and efficiency). A good starting point is asking developers to profile long-running processes or the most frequently running tasks to try to remove unnecessary code and checks, and to clean code portions that reflect years of modification and evolution. Reducing CPU cycles for each task will reduce power consumption and thermal dissipation, thus reducing carbon emissions. *Can we do more?*

We can start asking if the application is implemented with the right programming language. As you probably know, today there are a lot of programming languages, and over time, a lot of frameworks have been created, trying to shorten the **time to market (TTM)**, smoothening the implementation, and reducing the required programming skills. This is a strong and still valid pattern.

Instead of developing more skills to create cleaner and more efficient code, at some point we decided (this was probably cheaper) to keep the inefficient code and build it on frameworks or high-level programming languages, which allow a faster TTM with very little fuss.

Please consider that every layer you insert between the OS and your business application is a waste of efficiency that could be optimized or even removed: languages that are compiled to machine languages (such as **C**) usually perform way better than languages compiled to some kind of bytecode, which need a *process VM*, adding a layer of inefficiency.

A 2017 paper, *Energy Efficiency across Programming Languages* (`https://greenlab.di.uminho.pt/wp-content/uploads/2017/10/sleFinal.pdf`) compares different programming languages in various aspects and with standard algorithms, to identify the most energy-efficient language. Its results have been further updated with newer programming languages and their evolution in 2020: `https://sites.google.com/view/energy-efficiency-languages/`.

The result of these studies is very enlightening, but there are better-performing, greener programming languages that will allow our team to develop a highly *carbon-efficient* application. Of course, thinking that entire teams should change their coding skills is an impossible task, but maybe you can keep this in mind when you plan for the next project.

We've now seen how to plan for your application to be greener. The example we have worked on was gradually improved and optimized by substituting layers of the application with Azure PaaS services, where possible. This, by the way, has several additional advantages that we have explored throughout the book: from cost optimization to reducing your technical debt, to the final goal of reducing the application's carbon footprint. *But how can we measure its carbon impact?* This will be discussed in the next section.

Measuring and fostering sustainability

Now that we have learned what sustainable software engineering is, let's try to design a roadmap for carbon governance, just as we did with costs. Let's start with measuring our carbon impact, which we'll learn about in the next section.

The Microsoft Emissions Impact Dashboard

The first thing to do before any action is to be able to measure carbon efficiency. This is sometimes hard—for example, when you have hybrid applications—but if your workloads are all in Azure, then we have a dashboard that will help us out.

The **Emissions Impact Dashboard** is a free app that can be downloaded at `https://www.microsoft.com/en-us/sustainability/emissions-impact-dashboard`, provided you have Power BI Pro licenses and a valid billing contract (**Enterprise Agreement (EA)**, Microsoft Customer Agreement, or **Cloud Solution Provider (CSP)**). This will provide a view of your company's usage of cloud carbon emissions based on your subscriptions and service usage. This is calculated as a sum of compute, storage, and data transfer in the Azure cloud.

> **Important Note:**
>
> The Microsoft Emissions Impact Dashboard app is in the Preview stage. It requires you to have Power BI Pro. It also needs Admin access to the enrollment **identifier (ID)** or billing account ID of your Microsoft agreement to connect your billing data (this is also valid for **Microsoft Partner Agreement (MPA)**/CSP contracts). You may still use it with the sample data to check its content before connecting it to your billing data.

The following screenshot shows the default dashboard that is displayed when opening the app. As you can see, on this landing main page, you have a summary of your scope 1, 2, and 3 emissions, along with the projected end-of-year forecast and the history of your company's emissions:

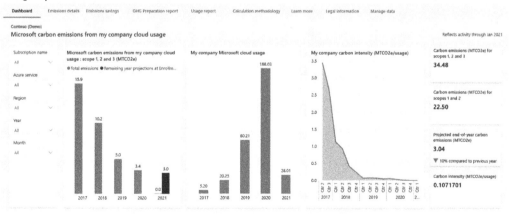

Figure 8.11 – The Microsoft Emissions Impact Dashboard (main page)

If you navigate to the next tab, as in the following screenshot, you have all the details of your emissions. These can be filtered by subscription and location (illustrated graphically), but most of all by Azure service. This view will give you a picture of which services have a higher carbon impact in your virtual data center:

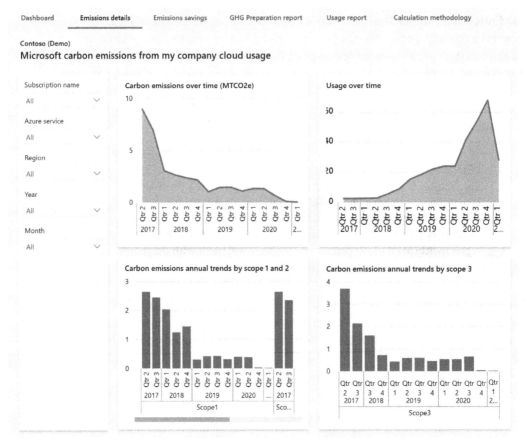

Figure 8.12 – The Microsoft Emissions Impact Dashboard (Emissions details): first half

The following screenshot shows the other half of the page illustrated in the preceding screenshot:

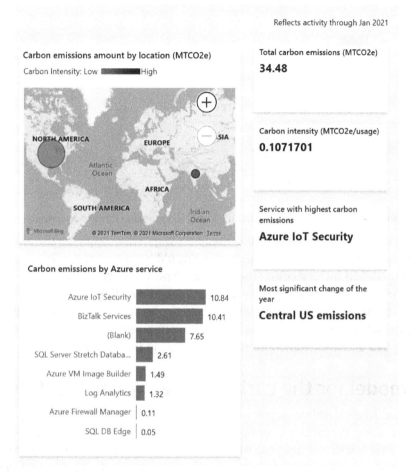

Figure 8.13 – The Microsoft Emissions Impact Dashboard (Emissions details): second half

The **Emissions savings** page, as shown in the next screenshot, is only related to an exercise of understanding the carbon emission savings of running your workloads in the cloud compared to an on-premises (albeit with different, customizable efficiency types) data center. This is not the type of saving we are targeting with our green software exercise, but it still holds very useful information for migrations:

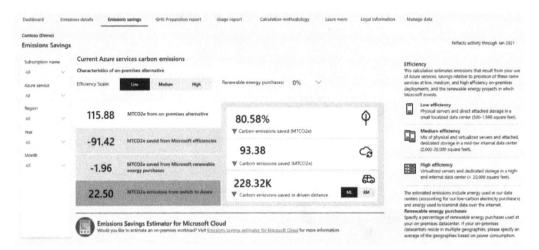

Figure 8.14 – The Microsoft Emissions Impact Dashboard (Emissions savings)

If you want to start tracking your carbon emission savings when applying the principles of sustainable software engineering, your best bet would be to export the information on the emission details by service and then create a report that will show your progress throughout the changes. We are not very far from Azure providing precise information on each service's emissions, but until then, we need to plan for practical steps and measurements and start logging changes to monitor our progress. We'll see how this can come together as a green software optimization process in the next section.

A data model for the carbon impact of applications

Having seen the Azure dashboard, you might think everything is laid out in front of you easily. But measuring the carbon intensity of an application that is not 100% on Azure, as you probably have learned by now, can be a difficult task, especially with the current onset and variety of hybrid and legacy architectures, often paired with native cloud *refactoring* to foster gradual migration.

Consider that the carbon impact goes beyond just the Azure data center. *Is the customer using ExpressRoute (which is a co-location (co-lo) partner connectivity provider) and other third-party dependencies through independent software vendors (ISVs)? What about edge devices?* The preceding reporting dashboards don't account for these ancillary services (at the time of writing) or provide application solutions for the customer and account for all its emissions. The carbon impact of an application should include many aspects that depend on its usage, as outlined here:

- The infrastructure (be it cloud or on-premises)
- The exchanged bandwidth

- The edge devices, wherever present
- The stored data along with its retention
- The end-user devices (along with their embedded carbon)

How can we easily measure the carbon consumption across all these components? Well, first of all, we can start by accepting that this can also be a relative measurement and not discrete. To be able to solve this puzzle, we should address measuring the carbon impact of an application by starting with the easy metrics that are widely available. This will allow us to measure the impact without sampling them one by one: as you might see, a customer with thousands of applications would make this exercise unachievable.

Let's start by picking one or two sample applications where we can concretely measure emissions. Then, we want to find a way to correlate them with standard tools. **Application performance management (APM)** tools are widely used and a common best practice to test how an application is behaving. The used metrics are very common, whereby we can find a subset of metrics that are common to the majority of APM products and can use them to correlate the model for our purpose: this will keep our model as open as possible, being available to the most used APMs.

For our scope, however, we will use Azure Application Insights: the main goal is to be able to take a fixed number of common application architectures (cloud-native, client/server, hybrid, and so on), initially manually measure their carbon usage, along with their APM metrics, and then produce a data model (possibly with **artificial intelligence (AI)**) that can do the following:

- Correlate the manual measurements to Application Insights metrics.
- Provide a forecast of carbon impact based on the output algorithm.

The benefits of this approach are the following:

- This can be a measurement-independent tool to get the application's carbon impact.
- This model can be even more precise than measurement, as it accounts for all the spikes in the usage of the application, not just compute or storage.
- It relies on widely used tools (such as **Azure Application Insights**) that everybody should already be using.

- Carbon metrics and forecasts can be embedded over time in the APM tools (by means of a *carbon dashboard* that will track the progress of the application through daily usage, optimization, and migration/refactoring).

- When a failure of the application occurs, this approach will allow a safe measurement of the carbon impact of failure, thus advocating for resiliency and other availability patterns.

- This might also be integrated with chaos engineering tools such as Azure Chaos Studio to intentionally introduce carbon-impacting events (for example, think of a spike in carbon emissions due to higher temperatures at the data center) and see how the application reacts by moving the workloads according to carbon impact: `https://azure.microsoft.com/en-us/services/chaos-studio/`.

The flowchart for assessing the carbon impact of your applications will look something like this:

Figure 8.15 – A model for measurement of an application's carbon impact

To summarize the preceding model, we can note the following:

1. We need to categorize our applications by type (that is, **content management system (CMS)**, web app, **internet of things (IoT)** app, workplace, and more) and their cost/carbon weight compared to each other.

2. Then, we choose a bunch of significant metrics from Azure Application Insights—for example, availability, send request time, processor time, HTTP request rate, process **input/output (I/O)** and CPU rate, server requests and request rate, and more.

3. We manually measure the carbon consumption of a sample set of our applications (by targeting emissions coming from CPU, memory, storage, and infrastructure usage), ideally one for each category.

4. We create an ML model that will correlate the chosen metrics to the carbon consumption of the sample apps.

5. We create a meaningful score (since it won't be a discrete measurement) that will help us correlate all the Application Insights metrics to the rest of the applications and store it for reporting over time.

As you might imagine, this approach is still experimental and there are many threads working on similar projects. The main idea is to generate awareness and find solutions that will trigger the creation of a standard measurement tool and application carbon score. Developers can then use them to enforce a governance process that tackles not just costs and savings for their applications but also their carbon impact, and continuously optimize both these aspects as a best practice and commitment to this planet.

Summary

With this final chapter, we have explored a recent discipline that is getting a lot of traction, due to a historical and political moment that requires studying sustainable software engineering.

This includes a series of techniques and principles that allow you to model your application with low carbon emissions options and architectural choices and even a possible data model to measure the carbon impact, to be able to track progress over time. The final intent is to merge the process of improving the carbon efficiency of our Azure applications into the overall cost governance process, thus having a complete set of best practices for an efficient, high-performing, and cost-effective digital transformation.

We are now at the end of our fascinating journey into the Azure cost governance process: the last thing you need to know is you should go back again to where we started and repeat the entire process.

We suggest this as the Azure tools, portal web page, and services will have changed over time, giving you more features, details, and granularity than what you had the last time. Also, your applications are not static either, and they will grow, change, be refactored, and be decommissioned, or you might have new stuff to consider compared to last time.

Our recommendation is that you immediately establish a recurring check that involves your cloud architects, your application managers, your cost controllers, and **cloud financial operations** (**FinOps**) people to regularly check your progress over time and incorporate any new operation. Once this has become a normal process in your cloud operations, you will be able to build additional knowledge into this domain by assessing how your set of workloads and applications behave over time.

If they are static and grow little with time, you might relax your schedule, but if you are in a fast-paced environment with lots of changes and additions, you need to focus on the automation of monitoring and policies to be able to govern faster and more accurately. *As with everything in life, cost governance requires discipline and care.* But doing it right will give you the satisfaction of bringing your Azure costs down sensibly (or at least controlling them) and making room for all the new, exciting, cloud-native apps that you will build on the **Azure** platform.

Questions

1. What are the main principles of sustainable software engineering?

2. What is demand shaping?

3. What is demand shifting?

Further reading

- GSF: https://greensoftware.foundation/

- Cloud Carbon Footprint open source tool: https://www.cloudcarbonfootprint.org/

- *Reducing Microsoft's carbon footprint by tracking internal Microsoft Azure usage*: https://www.microsoft.com/insidetrack/blog/reducing-microsofts-carbon-footprint-by-tracking-internal-microsoft-azure-usage/

- Microsoft technical blog on sustainable software: https://devblogs.microsoft.com/sustainable-software/

- *A Low Carbon Kubernetes Scheduler*: http://ceur-ws.org/Vol-2382/ICT4S2019_paper_28.pdf

Assessments

Chapter 1, Understanding Cloud Bills

Question 1

What is your billing type?

Answer

To check and validate your billing type, as per the documentation at `https://docs.microsoft.com/en-us/azure/cost-management-billing/manage/view-all-accounts#check-the-type-of-your-account`, you can check your type of billing directly from the **Azure Portal** | **Azure Cost Management** | **Properties** page.

Question 2

Can you provide a cost analysis with a daily stacked bar chart view of the last quarter?

Answer

Yes. To do so, open the **Azure Portal** | **Azure Cost Management** | **Cost Analysis** page, choose the *last quarter* view in the calendar selection, and then select `Granularity=daily`, `Column=stacked bar`.

Question 3

How can you export your billing information?

Answer

Billing information can be exported in four main ways, as follows:

- By manually exporting the **Cost Analysis** page in **comma-separated values** (**CSV**) or **Excel Binary File** (**XLS**) format

- By automatic export

- Via the **command-line interface** (**CLI**)

- Through the official **application programming interface** (**API**) provided by Microsoft (requires code to be developed)

Chapter 2, What Does Your Cloud Spending Look Like?

Question 1

What is a Meter category and a Meter subcategory? And what is the finest grained one that allows you to analyze specific dimensions?

Answer

They represent the *dimensions* of the usage of your Azure resources and can help you identify where you need to concentrate for optimization. The Meter subcategory is the finest grained one since you can have one or more subcategories for each Category.

Question 2

What is a tag, why is it so important, and how can you apply it with Azure resources?

Answer

A tag is a key/value string pair you can apply to every Azure resource using the **Tag** blade from the left menu in the Azure Portal. They are very useful for resource tracking and identification, as well as for operation, documentation, and cost control purposes.

Question 3

Are there resources in Azure that have only one Meter subcategory? How do you deal with optimizing those services?

Answer

Yes, of course—usually the simplest ones, or some basic tiers. You can still optimize such resources with right-sizing and, most importantly, by cleaning up unused items.

Question 4

Why do you need to change your server vision from pet to cattle?

Answer

Pets need attention, and when you lose your pet, everyone knows, and everyone is sad (or has an impact from it). With cattle, you can lose a member without basically noticing, and you can start thinking about maintaining the whole *outcome* of the cattle, replacing single members without being so sad.

Question 5

What are the native tools you can use to automate actions on Azure resources and cost savings?

Answer

The AZ CLI, the PowerShell CLI, API, and Bicep.

Chapter 3, Monitoring Costs

Question 1

Can you schedule an export of Azure cost information?

Answer

Yes. To schedule an export in the Azure Portal, navigate to **Cost Management | Cost Analysis | Exports** and follow the instructions to create a scheduled export.

Question 2

What is Azure Advisor?

Answer

Azure Advisor is a sort of assistant that will analyze your current Azure subscription(s) and come up with smart suggestions on many topics: costs, performance, reliability, security. The information is automatically calculated and loaded for you in a friendly dashboard, and you can create alert rules based on Advisor's suggestions.

Question 3

What is **Azure Hybrid Benefit** (**AHUB**)? Why is it important?

Answer

AHUB is a licensing benefit that will allow customers who already purchased Windows Server, **Structured Query Language** (**SQL**) licenses, and even RedHat and SUSE Linux subscriptions, to basically save on the cost of those on-premises licenses when running on Azure. The important thing to know is that you need to configure the AHUB Azure resources (either manually or via scripting), so that you don't have to pay for the license twice.

Question 4

How can you monitor your reservations' capacity and usage?

Answer

Reservations can be monitored via Power BI pre-packaged app reports or by manually using the information gathered in the Cost Analysis page of the Azure Portal..

Chapter 4, Planning for Cost Savings – Right Sizing

Question 1

Where can you get a list of **virtual machines** (**VMs**) to be right-sized?

Answer

The Azure Advisor **Cost** page has a specific entry for VMs that are not properly sized and can be customized by setting the percentage of idle **central processing units** (**CPUs**) that you wish to check and report upon. The default is 5%.

Question 2

Which types of managed disk storage can you choose from (cheapest to most expensive)?

Answer

Managed disks come in **Standard HDD**, **Standard SSD**, **Premium SSD**, and **UltraDisk** tiers (cheapest to most expensive).

Question 3

Can you downsize a VM with no downtime/impact?

Answer

If you are downsizing a single VM with no clustering or reliability option, then unfortunately you cannot downsize without downtime, since the resizing option will stop and start the VM.

Chapter 5, Planning for Cost Savings – Cleanup

Question 1

How do you know if you have unattached disks in your environment?

Answer

Search from the Azure portal within the list of managed disks for those with no owner (or an unattached status for unmanaged disks).

Question 2

How can you calculate the peering traffic of an application that shares the **virtual network (VNet)**?

Answer

You can calculate the peering traffic of an application that is sharing a VNET (for example, in a hub-and-spoke topology) by dividing the application's peering traffic by the sum of all the other applications' peering traffic meters.

Question 3

How can you delete an unused subscription?

Answer

You can delete an unused subscription in the Azure portal, provided you have the correct privileges to access the subscription and that you have made all the relevant checks on its effective usage.

Question 4

How can you get the costs of resources involved in a software migration project?

Answer

You can use the **Cost Management** page in the Azure portal, filtering by tags, selecting the impacted application **identifier** (**ID**) tag and the impacted landscape (Development, Testing, Production, Upg_Prj_*, and so on), and grouping by Meter category, Meter subcategory, or Resource.

Question 5

How can you split the costs of a shared resource?

Answer

You initially need to define a driver that could be storage-based (dedicated disks) or CPU-based (CPU usage for each running workload). Then, you need to identify which costs are shared (for example, VM compute cost, **operating system** (**OS**) disk, snapshot, or network) and what are the dedicated costs, if any (for example, dedicated storage). At this point, you have shared and dedicated total costs and the percentage to split the shared ones.

Chapter 6, Planning for Cost Savings – Reservations

Question 1

How can you know whether a VM reservation is worth it?

Answer

You can calculate this by dividing the monthly reserved price by the hourly pay-as-you-go, which is the minimum number of hours for which a VM should run for the reservation to be worth it..

Question 2

What is the **Amortized view**?

Answer

The **Amortized view** will split the reservation quota of each resource into your daily billing.

Question 3

Why does your cost controller need to know about reservations?

Answer

The reservation process should be carefully explained and shared with your cost controller for the following reasons:

- Some companies may consider reservations as **capital expenditures (CAPEX)** and not **operating expenses (OPEX)**, so your budget should reflect this point.

- Some companies need you to re-discount the monthly competence over the years, and you need to track down exchanges and monetary commitment.

Chapter 7, Application Performance and Cloud Cost

Question 1

What are the limits of infrastructure-only optimization, and why do you need to work with the app and dev team for any application optimization?

Answer

With an infrastructure-only optimization approach, you can right-size, implement a scheduled or on-demand on-off automation, or reserve capacity. Working with developers, you can optimize workloads to better use Azure resources and services and have a cheaper and better-performing application, on top of an optimized infrastructure.

Question 2

Why can database tuning be important for cost saving?

Answer

Because even if you already optimized your entire infrastructure, your database queries might not be optimal and force you to over allocate both compute and storage resources. Bringing the database requirements down can have a direct impact on your cloud savings.

Question 3

What is technical debt?

Answer

Technical debt is the need to refactor and improve your application, be it legacy or cloud-native, by estimating roughly the time needed to fix things that are not right and change architectural patterns that have evolved: it is measured in man-days of effort to fix things, or in versions missing to reach the latest valid release.

Question 4

Why is Application Insights a key element of application optimization?

Answer

Having an **application performance management (APM)** service such as Application Insights is critical to understanding whether your application has substantial architectural issues, but also to spot that little hidden problem that might be found by one user after a million-page hits but that will trigger a devastating number of faults.

Chapter 8, Sustainable Applications and Architectural Patterns

Question 1

What are the main principles of sustainable software engineering?

Answer

The main principles of sustainable software engineering are carbon, electricity, carbon intensity, embedded carbon, energy proportionality, networking, demand shifting and shaping, monitoring, and optimization.

Question 2

What is demand shaping?

Answer

Demand shaping is a technique that allows you to shape your application according to specific parameters or the end user's choice to save on carbon emissions.

Question 3

What is demand shifting?

Answer

Demand shifting is a technique that allows your cloud application, based on specific conditions, to move workloads to regions that have a lower carbon impact, thus saving on emissions.

Index

C

V

W

Z

`Packt.com`

Subscribe to our online digital library for full access to over 7,000 books and videos, as well as industry leading tools to help you plan your personal development and advance your career. For more information, please visit our website.

Why subscribe?

- Spend less time learning and more time coding with practical eBooks and Videos from over 4,000 industry professionals

- Improve your learning with Skill Plans built especially for you

- Get a free eBook or video every month

- Fully searchable for easy access to vital information

- Copy and paste, print, and bookmark content

Did you know that Packt offers eBook versions of every book published, with PDF and ePub files available? You can upgrade to the eBook version at `packt.com` and as a print book customer, you are entitled to a discount on the eBook copy. Get in touch with us at `customercare@packtpub.com` for more details.

At `www.packt.com`, you can also read a collection of free technical articles, sign up for a range of free newsletters, and receive exclusive discounts and offers on Packt books and eBooks.

Other Books You May Enjoy

If you enjoyed this book, you may be interested in these other books by Packt:

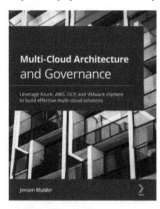

Multi-Cloud Architecture and Governance

Jeroen Mulder

ISBN: 978-1-80020-319-8

- Get to grips with the core functions of multiple cloud platforms
- Deploy, automate, and secure different cloud solutions
- Design network strategy and get to grips with identity and access management for multi-cloud
- Design a landing zone spanning multiple cloud platforms
- Use automation, monitoring, and management tools for multi-cloud
- Understand multi-cloud management with the principles of BaseOps, FinOps, SecOps, and DevOps
- Define multi-cloud security policies and use cloud security tools
- Test, integrate, deploy, and release using multi-cloud CI/CD pipelines

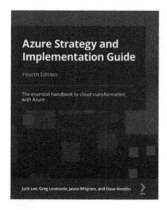

Azure Strategy and Implementation Guide - Fourth Edition

Jack Lee , Greg Leonardo , Jason Milgram, and Dave Rendón

ISBN: 978-1-80107-797-2

- Understand core Azure infrastructure technologies and solutions
- Carry out detailed planning for migrating applications to the cloud with Azure
- Deploy and run Azure infrastructure services
- Define roles and responsibilities in DevOps
- Get a firm grip on Azure security fundamentals
- Carry out cost optimization in Azure

Packt is searching for authors like you

If you're interested in becoming an author for Packt, please visit authors. packtpub.com and apply today. We have worked with thousands of developers and tech professionals, just like you, to help them share their insight with the global tech community. You can make a general application, apply for a specific hot topic that we are recruiting an author for, or submit your own idea.

Share Your Thoughts

Now you've finished *The Road to Azure Cost Governance*, we'd love to hear your thoughts! Scan the QR code below to go straight to the Amazon review page for this book and share your feedback or leave a review on the site that you purchased it from.

https://packt.link/r/1-803-24644-8

Your review is important to us and the tech community and will help us make sure we're delivering excellent quality content.

www.ingramcontent.com/pod-product-compliance
Lightning Source LLC
Chambersburg PA
CBHW062107050326
40690CB00016B/3238